LULU PRESS

CONSULTANT MEDICAL INTERVIEWS

A Comprehensive Guide to Consultant Interviews

Updated and Revised 5th Edition(2015)

Keeping you ahead, always

www.consultantmedicalinterview.com

ISBN 978-1-326-17649-5

Table of Contents

Dedicated to the memory of my father who gave me more than I realised.

The author is a consultant physician in the NHS and wishes to remain anonymous.

Consultant Medical Interview

The fourth edition of the book has been an overwhelming success. Many thanks for your words of support and success stories. The NHS is undergoing one of the most radical changes in the history of health service. The fifth edition has been comprehensively revised and updated to reflect these radical changes. We have also removed some of the non-relevant sections from the book.

As before, the book is designed to inform, update and prepare you comprehensively for the consultant interview.

I believe that interview courses can only fine-tune your preparation. They aren't designed to organise your preparative thinking from the bottom up. I am convinced that interview courses can only help if you have done your preparations first. The information presented here will help you achieve that.

This book aims to kick start your preparations for the interview. It will aim to provide concise information and content for your interview preparation. All the answers are indicative only. The information is useful for any medical interview but concentrates on the consultant interview.

Good luck!

Preparing to apply for the Consultant Job

When should you start looking for the consultant job?

Start early, at least 12 months before your CCT date. This will give you sufficient time to improve your CV depending on the job you want. Think about:

- Where would you like to work?
- Who would know about retirements and consultant expansion in your specialty?
- Visit individual trust sites of interest
- Talk with your programme director

When can you apply?

You can apply 6 months before your CCT date. Register with www.jobs.nhs.uk early. The jobs of interest will be emailed to you as soon as they become available. Also visit the BMJ careers website weekly.

Choosing the hospital to apply to

You may wish to consider the following before shortlisting a hospital for your application

- Remember to choose your hospital carefully. It is generally a job for life.
- Be careful about a job in the hospital where you have worked as a senior trainee. You will always be a trainee in the eyes of your consultant colleagues.
- Look at all the information available with the application form. This may help in your decision making.
- You can find out more about the hospital from www.drfosterhealth.co.uk/hospital-guide or from the latest inspection reports on the Care Quality Commission website- www.cqc.org.uk.

- Remember it is not all about the hospital. The geography is important too! You may want to look at recreational facilities, schools, houses and their cost (can you afford it?).

What do you do once you have decided to apply for a job?

- Phone the contact in the advert

- Enquire about what they want, even if it is in the advert, so that you can emphasise this in the CV

- Ask if you can visit. A pre-application visit is not obligatory, but it improves your chances of being shortlisted.

- Pre application visit:
  - It helps you decide whether to apply and helps you get shortlisted. At this visit, you could assess the position: what are colleagues like? What opportunities are there? What problems exist? Is it what you want? You could also discuss with specialist trainees to get some more information or a feel for the place.
  - Who do you meet at a pre-application visit? Meet potential colleagues—your specialty consultant colleagues and related specialty consultants like surgeons, pathologists, radiologists etc.
  - Check out the people and vibes, talk to juniors, secretaries etc.
  - Check out the facilities
  - Consider taking a partner to check out the area – an unhappy partner will make for a miserable job!

Reality check- Is there a 'local' candidate? Local trainees usually have an advantage.

So to summarise:

What job do you want?

Ideally decide early, so that you can tailor your CV

- DGH 'v' Teaching hospital
- Large 'v' small (generalist /specialist)
- Type of facilities etc.
- Region / amenities
- Researching the job
- Advert, job description, local knowledge, inside information

Consider

- Potential to support / develop your interests
- Colleagues (particularly young ones)
- Support services – radiology, pathology, surgeons
- Mix of acute medicine/surgery and your specialty

NHS job application

Most of the current jobs need an online application using the generic NHS application form. There is a word limit for each section. It is thus important that you maximise your strengths in the application using facts.

Top Tips

- Write your application with the person specification in front of you

- Print out the form to see how it looks, before clicking 'submit'

Remember, the forms are anonymised. Hence, your personal details will not be visible to the person doing the shortlisting.

Example

Describe your experience of clinical audit. (max. 150 words)

I have undertaken and presented eight audit projects during my registrar training:

List your audit projects. Write a line or two detailing the key outcomes and any changes you initiated as a result of these outcomes. Try and complete at least one audit cycle. This demonstrates that you believe in the principles of audit and that it was not a 'tick box' exercise.

Describe your relevant teaching experience. (Max 150 words)

- I have been an avid teacher ever since my postgraduate days

- I have done the Postgraduate Certificate in Medical Education.

- I am a recognised teacher for the MBBS degree programme at Balamory University.

- I am involved in teaching and assessing medical students from the University of Balamory. This involves both non-clinical tutorial sessions and also ward-based clinical teaching during the student's attachment to the (specialty) department.

- I regularly teach junior doctors preparing for the membership examination. This teaching incorporates an objective-structured clinical examination and a communication skills practice.

- I have initiated and organised a departmental teaching programme for junior doctors in (specialty) at Balamory hospital.

- I have attended the Teaching the Teachers for Specialist registrars.

(Tips: Again, you mention facts like PG Cert, teaching the teachers course etc. to demonstrate your passion for teaching. You could mention a student survey and the feedback received to demonstrate that you are a good teacher)

Details of your most relevant research work and publications in peer-reviewed journals. (max. 500 words)

I have been involved in clinical research throughout my postgraduate career. My research projects include: [Enumerate your research projects and your role in it].

a. Study I.

I was the Principal Investigator for the study. I was responsible for the Ethics and Research & Development approval. I initiated, wrote and managed the study. This was a single centre study and recruited 18 patients.

b. Study II.

A project submitted to and accepted by Balamory University for the award of MD degree. I was the Principal Investigator for the study. I was responsible for recruiting and coordinating the care of patients admitted with [details]. All the patients recruited were investigated with a defined study protocol.

Peer reviewed publications

List your peer-reviewed publications chronologically in Vancouver style. Put your name in bold.

Papers currently submitted for peer-review

You could use a separate heading for papers currently under consideration for publication.

Give examples of your approach to working in a team. (max. 150 words)

I have done 360 degree appraisals twice and both have been very positive about my team-working spirit.

My participation in the national audit revealed poor practice in the use of (drug). I raised the issue at the local clinical governance meeting. I subsequently produced a local guideline for the use of (drug) after collaborating with all the stakeholders like physicians, biochemists, pharmacists and ward nurses. This was a valuable lesson in initiating and leading a change by consensus using an evidence-based approach.

As principal investigator for my research project, I successfully led a team of scientists, doctors and nurses to finish the project. On reflection, I realised the value of prioritisation, planning, mutual respect and communication in the successful completion of the research on time.

Presently, I organise a regular departmental teaching programme for junior doctors at Balamory hospital.

(Tip- Again <u>mention facts</u> rather than a long waffle about how great a team player you are!)

Please explain your areas of clinical skill and competence relevant to this post. (max. 150 words) (Tip: Sit down with the person specification form)

- I have gained extensive experience in luminal gastroenterology, hepatology and nutrition.

- I have performed in excess of 1000 gastroscopies. I am competent in the therapeutic procedures of oesophageal dilatation and stenting, endoscopic therapy of variceal and non-variceal bleeds and insertion of PEG tubes.

- I have done in excess of 500 colonoscopies and I have extensive experience of snare polypectomy, removal of sessile polyps and terminal ileoscopy. My caecal intubation rate is more than 90% with an adenoma detection rate of 21%.

- I am competent at reporting capsule endoscopies and have double-reported 25 capsule endoscopies with excellent agreement.

- I have had 4 years of resident general acute medicine experience as a registrar. This has involved responsibility for all acute admissions, including coronary, respiratory and intensive care patients under the care of the physicians. I am fully competent in all general medical practical procedures.

Please provide any other supporting information that you think may be helpful, or that is requested in the Person Specification. Please ensure that this does not contain any duplicate information already provided elsewhere in the application form or any personal details. (max. 500 words) (Tip: Sit down with the person specification form)

Brief description of your duties and responsibilities (4000 characters)

I have been working as a Specialist Registrar in the Balamory LETB (previously deanery) since 2004. During this period, I have taken every opportunity to gain further knowledge and training in the fields of clinical work, teaching, audits, clinical governance, management activities and administration.

My post involves assisting with the provision of the clinical services for elective and emergency medical admissions and endoscopic procedures. I participate in resident general acute medicine on-call on a 1:12 rota with prospective cover. Supervision, assistance or advice appropriate to clinical need is available at all times.

My weekly commitment includes 2-3 outpatient clinics, 3 independent endoscopy lists and 2-3 ward rounds.

My training as a SpR has given me thorough systematic exposure to general gastroenterology, hepatology, nutrition and endoscopy.

I have had a varied experience in general gastroenterology including the in- and out- patient management of oesophageal diseases, peptic ulcer disease, inflammatory bowel disease and pancreatic diseases.

My training in hepatology was augmented by a regular monthly liver clinic, joint HIV hepatitis clinic and liver transplant assessment and review clinics. Participation in monthly liver MDT involving transplant surgeons, transplant physicians, pathologists and the radiologists provided me with valuable experience in dealing with difficult management issues in hepatology.

My training has been enhanced by my role as part of the nutrition team (for the last 1.5 years) where I have been asked to give specialist nutrition advice to other specialties besides assessing patients for PEG, and a regular review of patients with home TPN and reviewing in-patients needing nutrition support.

I had regular competency-based endoscopy training and assessments as per JAG guidelines.

During the final two years of my training, I have been taking up increasing responsibilities in the clinical area. I supervise the core trainee and the junior registrar. I have been doing independent colonoscopy lists and have taken an increasing role in the provision of emergency endoscopies. I take up the leadership position in the management of complex cases. I co-operate and communicate with colleagues and recognise these as essential skills in being an efficient team player.

I regularly participate in the gastroenterology radiology and pathology meetings besides the cancer MDT.

Continuing Education and Professional Development

Meetings

I regularly attend departmental audit, morbidity and mortality meetings and journal clubs. I have also attended several regional and national meetings.

Teaching and Training

All hospitals within our training region provide a varied weekly teaching programme. As well as one-to-one teaching in the endoscopy suite, there are regular formal training days across the sites in both gastroenterology and general medicine.

Management experience

I have gained a useful understanding of how things are run, how change occurs, and how to influence that change. I realise that the partnership between doctors and managers is inevitably tricky in a resource-limited NHS. I am, however, determined to do my best for the patients while actively working with the management.

- I am the SpR representative for the development of Emergency care pathway at Balamory Hospital.

- I regularly attend the directorate meetings and participate in the discussions about service organization and delivery, individual workloads, waiting lists etc.

- I have initiated and organised a departmental teaching programme for junior doctors in (specialty) at Balamory hospital.

- I organised and coordinated the clinical part of the MRCP exam at the Northern University hospital

- I developed the trust guidelines for the use of (drug) in (clinical condition). This was a valuable experience in clinical governance.

- The management development course has provided me with valuable practical skills and has put me on a steep learning curve.

Structure of Curriculum vitae

A top CV should be

- Clear
- Open and honest
- Layout is paramount & must be logical and easy to follow
- Highlight good jobs and experience
- Concise- no waffle
- Use domains and bullet points
- Tailor your CV based on the person specification
- Always remember that quality is more important than quantity

Structure of a CV

Page 0	Name, qualifications & job application details
Page 1	Personal details (Name, contact details, date of birth, GMC status, CCT completion date) and qualifications
Page 2	Career details- List all the jobs chronologically
Page 3	Specialty experience
Page 4	Research experience
Page 5	Audits
Page 6	Teaching experience
Page 7	Management experience
Page 8	Prizes and awards
Page 9	Computer and Language skills
Page 10	Personal interests
Page 11	Courses and meetings attended
Page 12	Society and Professional memberships

Tips

- Use the same font throughout. Arial/Verdana 14 bold for headings and size 12 for content
- Don't use CAPITALS for headings. Bold or increased size is more professional
- Use a footer to display your name and page number
- Use good quality white/cream A4 paper (100gsm)
- 2.5cm margins all around.
- Use tables appropriately. Don't use them if you are not confident.
- Use a single staple in the left corner. Alternatively you can use professional presentation folders.
- Get feedback from at least 2 consultants
- A consultant's CV generally has 12-16 pages
- Spell check!
- Be wary of non-medical website advice on CV writing. They do not always apply to medical CVs
- Be thorough but concise in your descriptions. Do not fill your CV with unnecessary words just to make it longer. It will reflect badly on your ability to express yourself in a clear and concise manner.

Further reading

Writing a winning CV. Sam McErin: http://tinyurl.com/5wxonml

Growing your CV: Elitham Turya: http://tinyurl.com/649r5ko

Writing a CV: http://tinyurl.com/27uonls

Dr. Name

MBBS, MD, MRCP

Curriculum Vitae

Application for the post of Consultant Acute Medicine at

Balamory Hospital

Nov 2012

Personal Details

Name:
Date of birth:
Contact details:

Address:
Email:
Home tel:
Mobile tel:

GMC number:
CCT date:

Qualifications

Jun 2011	Fellow of the Higher Education Academy
May 2010	Postgraduate Certificate in Medical Education. Balamory University, UK
Jun 2008	MRCP
	Royal College of Physicians, UK
May 2006	MD
	Balamory University
Dec 2001	MBBS
	Balamory University

Experience

Present Appointment

Apr 2009 to present

Specialist Registrar in General Medicine and (specialty)

Name and address of hospital

Supervising consultants

Previous Appointments

Apr 2008 to Mar 2009

Specialist Registrar in General Medicine and (Specialty)

Name and address of hospital

Supervising consultants

List all your jobs from PRHO onwards

Overseas Appointments

Write overseas experience if any under a separate heading.

Professional Training

Gastroenterology Training

Clinical

- I have gained extensive experience in gastroenterology over the last five years. Experience in general gastroenterology includes the in- and out- patient management of oesophageal diseases, peptic ulcer disease, inflammatory bowel disease and pancreatic diseases, and in hepatology, the management of acute and chronic liver disease with their associated complications.

- My training in hepatology was augmented by regular liver clinics, joint HIV hepatitis clinics and liver transplant assessments and review clinics. Participation in monthly liver MDTs involving transplant surgeons, transplant physicians, pathologists and the radiologists provided me with valuable experience in dealing with difficult issues in hepatology.

- My training in functional bowel diseases was consolidated by a weekly IBS clinic led by a consultant and supported by a specialist nurse.

- I have developed a keen interest in nutrition. I have been a part of the multidisciplinary nutrition support team for the last year and a half. The team plays a vital role in providing specialist nutrition advice to colleagues with patients who have difficult nutritional problems. The team reviews and assesses all patients undergoing a gastrostomy tube insertion (PEG or RIG) and this has significantly reduced the mortality and morbidity of this procedure when compared with the national average. The team regularly reviews all patients on TPN, reviewing the indications and looking out for potential complications and treating any that may arise early and efficiently.

Endoscopy Training

Upper GI endoscopy

I have extensive experience of both diagnostic and therapeutic upper-gastrointestinal procedures. I have performed in excess of 1000 gastroscopies. My therapeutic experience includes:

- Oesophageal dilatation for benign strictures, achalasia and malignant disease
- Insertion of self-expanding metal oesophageal endoprostheses
- Insertion of percutaneous endoscopic gastrostomy (PEG) feeding tubes
- Insertion of naso-jejunal tubes
- Endoscopic therapy for bleeding peptic ulcers including clips and APC
- Endoscopic therapy for oesophageal and gastric varices including band ligation, sclerotherapy and the use of other injectates.
- Use of LASER for oesophageal tumours.

Lower GI endoscopy

I have performed in excess of 500 colonoscopies. I have extensive experience of rigid and flexible sigmoidoscopy and also diagnostic and therapeutic colonoscopy including snare polypectomy, techniques for the removal of sessile polyps, hot biopsy, argon plasma coagulation of angiodysplasia / irradiation telangiectasia and terminal ileoscopy. My caecal intubation rate was more than 90% in the last year with an adenoma detection rate of 20%.

Video Capsule Endoscopy

I am competent at reporting capsule endoscopies and have double-reported twenty six capsule endoscopies so far with excellent agreement in the detection of both positive and negative findings including the therapeutic suggestions.

General Medical Training

I have had 4 years of resident general acute medicine experience as a registrar on a 1:7 to 1:12 ratio with an average number of admissions varying from 20 to 60. This has involved responsibility for all 'unselected' acute medical admissions, including coronary and intensive care patients under the care of the physicians. Previous appointments have afforded widespread experience in general internal medical practice with specialised experience in cardiology, coronary care units, medical high dependency units, nephrology, neurology and respiratory medicine as a core trainee. During my overseas training, I gained experience in the management of various tropical and infectious diseases.

In addition, I am fully competent in a variety of practical procedures including:

- Insertion of central venous catheters
- Pleural aspiration and biopsy
- Insertion of chest drains
- Lumbar puncture
- Insertion of temporary pacing wires
- Percutaneous liver biopsy
- Ascitic paracentesis- diagnostic and therapeutic
- Use of Sengstaken-Blakemore tubes

Research Experience

Study 1

Name of the hospital 2007-2009

I was the Principal Investigator for the study. I was responsible for the Ethics and Research & Development approval. I initiated, wrote and managed the study. This was a single centre study and recruited 18 patients.

Brief background and results of the study (5-6 lines maximum)

Study 2

Audits

I recognise the value and importance of audits in medical practice. I have participated in audits throughout my career. Projects that I have carried out include:

List the title and date of audits. Short summary of each audit with reason behind it, your role, key outcomes and actions taken as a result (2-3 lines)

Audit 1- Name of hospital 2011

Key outcomes of the audit and actions taken (for e.g. I produced local trust guidelines).

Audit 2- Name of the hospital 2009

Teaching Experience

I have always enjoyed teaching. I did the postgraduate certificate in medical education by distance learning recently to further enhance my teaching skills. The course involved 20 units dealing with the key issues of Curriculum Development, Assessment, Teaching and Learning, Mentoring and Student Support and Quality Assurance. The course has helped me in becoming a more effective teacher.

Undergraduate- I am a recognised teacher for the MBBS degree programme at the University of Balamory and am involved in teaching and assessing medical students. This involves both non clinical tutorial sessions and also ward-based clinical teaching during the student's attachment to the Gastroenterology department. These teaching sessions follow a set syllabus and utilise various methods that I am familiar with, such as ward-based teaching and problem based learning.

Postgraduate-On a postgraduate level, I regularly teach junior doctors preparing for the membership examination. This teaching incorporates a multiple choice question practice, an objective structured clinical examination and a communication skills practice. I have initiated and led a current teaching programme for junior doctors attached to the (specialty) firm.

Management Experience

I have gained a useful understanding of how things are run, how change occurs, and how to influence that change. I realise that the partnership between doctors and managers is inevitably tricky in a resource-limited NHS. I am, however, determined to do my best for the patients while actively working with the management.

- I regularly participate in the directorate meetings. These meetings involve discussions about service organization and delivery, individual workloads, waiting lists etc.

- I have initiated and organise a departmental teaching programme for junior doctors in (specialty) at (hospital).

- I organised and coordinated the clinical part of the MRCP exam at the Balamory hospital

- I developed the trust guidelines for the management of (clinical condition). This was a valuable experience in clinical governance.

- The management development course has provided me with valuable practical skills and has put me on a steep learning curve.

Computer and Language Skills

I am fully conversant in the use of computers and I am proficient in the use of Word, Excel, Access and PowerPoint to carry out various non-clinical duties.

I am fluent in the use of the English and French languages.

Personal Interests

I enjoy playing cricket, table tennis and chess although I don't get a lot of time to indulge myself. I have a lovely three year old daughter. I enjoy spending time with her. I also enjoy the cinema, music and reading thrillers. I have a keen interest in the development and use of web 2.0 tools in medical education.

Courses and Meetings Attended

Gastroenterology

Sep 200X	Therapeutic GI Endoscopy course, London
Feb 200X	Intermediate skills in Colonoscopy, London
Nov 200X	The European Capsule Endoscopy Castle Course
Sep 200X	The Leeds Course in Clinical Nutrition
Sep 200X	Foundation course in Colonoscopy, London
Jun 200X	Foundation course in Endoscopy, London

General (internal) Medicine

Jun 200X	Management Development Programme, London
Dec 200X	Teaching the Teachers, London
Oct 200X	Foundation Programme Assessment Training, London
Mar 200X	Problem Based Learning Training, London
Jan 200X	Introduction to Research methods, Newcastle
Oct 200X	Advanced Life Support Newcastle
Sep 200X	Good Clinical Practice & the EU Trials Directive

Society and Professional Membership

British Society of Gastroenterology

American Gastroenterological Association

British Association for Parenteral and Enteral Nutrition

British Medical Association

Medical Defence Union

Collegiate Member of Royal college of Physicians of London

Fellow of the Higher Education Academy

Publications

Peer-reviewed

Please provide the references in Vancouver style.

Papers currently submitted for peer-review

Please provide the references in Vancouver style.

Abstracts arising from presentations to learned societies

Please provide the references in Vancouver style.

Abstracts currently submitted

Future Aspirations

I believe in excellence, innovation and reliability. I am passionate about (special attribute required in the job e.g. education, endoscopy etc.).

Add a line or two in line with the job specification.

Referees

CV Example 2

Sample Respiratory CV

Dr. Name

MBBS, DCH, MRCPCH

Curriculum Vitae

Application for the post of Consultant in Paediatrics with a special interest in diabetes

At

(Application Reference ......)

Personal Details

Name:

Address: Address

                                Email:

                                Home tel:

                                Mobile:

Date of birth:

G.M.C number:

Date of CCT completion:

Medical Qualifications

MRCPCH   Oct 20XX

M.B.B.S   Sept. 19XX

Other Qualifications

APLS (provider):   Balamory Hospital, 20XX

      (provider):   Royal Infirmary; Oct. 20XX

NLS (Selected as instructor):   Australian Hospital, 20XX

      (provider):   South Hospital; Aug. 20XX

Awards

2nd Certificate of Honour in Obstetrics and Gynaecology Gold Medal Examination (19XX).

Positions Held

Specialist Registrar

Mar XX- present	General Paediatrics, Diabetes & Endocrinology
	Hospital
Sept 'XX - Mar 'XX	Paediatric Diabetes and Endocrinology (6m)
	Hospital
Mar 'XX- Sep 'XX	Diabetes, Endocrinology & General Paediatrics (6m)
	Hospital
Mar 'XX - Mar 'XX	Neonatology, tertiary level
	Hospital
Sept 'XX – Mar 'XX	Community Paediatrics, on call- Paediatrics (6m)
	Hospital
Mar 'XX – Sept 'XX	Neonatology, tertiary level
	Hospital
Sept 'XX – Mar 'XX	General and Respiratory Paediatrics (6m)
	Hospital
Mar 'XX- Sept 'XX (LAT)	PICU (3m) and General Paediatrics (3m)
	Hospital

Senior SHO (SpR responsibility)

List your senior SHO jobs if any

Senior House Officer/Core training

List all your SHO/ core training jobs

P.R.H.O/Foundation jobs

List your PRHO/FY jobs

Details of Clinical Experience

General Paediatrics

- Competent in resuscitation of seriously ill children using APLS guidelines.

- Competent in management of children who require high-dependency care and stabilisation of children prior to transfer to PICU. Three-month experience of working in PICU.

- Competent in management of children with a wide range of common childhood diseases.

- Experience in planning discharge of complex long stay patients.

- Regular unsupervised teaching ward round with SHO/core trainees.

- Supervision of SHO/core trainees and junior registrars in their clinical duties including technical paediatric procedures.

- Competent in common paediatric procedures including long line insertion, lumbar puncture, intraosseous cannulation, pleural fluid aspiration. Very good experience in procedures including chest drain insertion, bone marrow aspiration, chemotherapeutic drug administration via intravenous and intrathecal routes and liver biopsy.

- Regular experience of a children outpatient clinic in general paediatrics and sub-specialty.

- Six months experience as registrar with special interest in Respiratory Paediatrics. Gained experience in asthma and cystic fibrosis clinics.

Child protection

- Competent in child protection procedures and medical examinations.

- Participated in a separate on-call rota for child protection at the University Hospital of North Staffordshire and at Birmingham Heartlands Hospital.
- Attended child protection courses and case conferences.

Neonates

- Competent in all aspects of modern neonatal care.
- Competent in resuscitation of very premature and sick neonates.
- NLS Instructor. Supervise SHO/core trainees and nurse practitioners at resuscitation.
- Competent in the use of conventional ventilation, high frequency oscillation ventilation and inhaled nitric oxide therapy.
- Competent in neonatal intensive care skills such as endotracheal intubation, umbilical catheterisation, insertion of long lines and chest drain. Teach SHO/core trainees and nurse practitioners these skills.
- Competent in cranial ultrasound scanning. Successfully completed 4 module course in cranial ultrasound scans.
- Competent in the management of babies with congenital anomalies.
- Competent in antenatal counselling of parents of premature babies.
- Gained very good experience in the process of withdrawal of intensive care from dying babies.
- Competent in transportation of sick newborns.

Paediatric Diabetes:

- About to complete two year experience in paediatric diabetes including six month experience at tertiary level.

- Competent in management of children with Type 1 diabetes in different age groups and in the use of different insulin regimens.

- Competent in the process of annual review of children with diabetes.

- Competent in assessment and management of DKA.

- Competent in managing emergencies like hypoglycaemia and missed or incorrect insulin doses. Regular supporter of specialist nurses with these emergencies.

- Competent in management of diabetes at the time of surgery and other illness.

- Journal club presentation, experience of audit in relation to diabetes.

- Participation in teaching of nurses in relation to diabetes.

- Gained good experience in the process of transition and attended joint clinics with consultant in adult diabetes.

- Gained good experience in managing children with type II diabetes and diabetes in children with cystic fibrosis.

Paediatric Endocrinology:

- About to complete a two year experience in paediatric endocrinology including six months at tertiary level.

- Competent in assessment and management of common paediatric endocrine disorders involving growth, puberty, thyroid gland and adrenal gland disorders.

- Competent in conducting and interpreting various endocrine investigations.

- Gained good experience in management of other endocrine disorders including ambiguous genitalia, obesity and disorder of calcium metabolism.
- Attendance and presentations at regional endocrine group meetings.

Community Paediatrics

- Experience in developmental assessment and ADHD clinics.
- Running of school clinics and child health clinics.
- Regular educational statement assessments.
- Attended outreach clinics and special school clinics.
- Participated in the multi-disciplinary assessment of children at the child development centre.

Courses and Training Programmes Attended

- Managing poor performance in Junior Doctors, Hospital and date
- European Society for Paediatric Endocrinology, Annual Meeting, Istanbul. Sep 2008
- Teaching the Teachers for Consultant Paediatricians and Senior SpRs (appraisal and assessment), place, date
- Stepping up to your consultant role (Management Awareness) course, May 20XX,
- Advanced Course in Paediatrics & Adolescent Diabetes, April 20XX, Hospital
- Advanced course in Paediatric Bone and calcium metabolism, Feb 20XX, Hospital
- Clinical Governance and Patients Safety training, Oct 20XX. Place

- Generic Instructor Course for NLS, Date, Hospital
- Neonatal Cranial Ultrasound Course, Date, Hospital
- Endocrinology for the General Paediatrician, Date, Hospital
- Cardiology in Neonates and Infants, Date, Hospital
- Dealing with Problem Colleagues, Date, Hospital
- Child Protection Training for Senior Clinicians, Date, Hospital
- Workshop on Diagnostic Methods, Date, Hospital
- Dermatology Course for Paediatricians, Date, Hospital
- Management awareness course for specialist registrars, Date, Hospital
- Evidence based medicine and critical appraisal of literature, Date, Hospital
- An Overview of Communication skills, Date, Hospital
- Day meeting on cystic fibrosis, Date, Hospital
- Specialist Registrar Teacher's Training, Date, Hospital
- Neonatal and Paediatric Ventilation Course, Date, Hospital
- Paediatric and infant critical care transport course, Date, Hospital
- Child Protection Training course, level II, Date, Hospital
- Child Protection Training course, level I, Date, Hospital
- Cranial ultrasound in the Newborn, Date, Hospital
- Paediatric Oncology Workshop, Date, Hospital
- National Paediatric Pulmonology conference and workshop, Date, Hospital

Professional Affiliations:

- Member, Diabetes UK
- Member, BSPED: British Society for Paediatric Endocrinology and Diabetes
- Member, Royal college of Paediatrics and Child Health
- Member, Medical Protection Society

Audits

List your audits. Mention the key findings and the actions you took to change practice as a result of the audit.

Teaching Experience

I have been involved with teaching at all levels and comfortably use the different methods of teaching. My teaching experience includes:

- Instructor on NLS course.
- Examiner for final year Paediatric OSCE
- Teaching at regional training programme for paediatric SpRs.
- Teaching at regional training programme for senior house officers.
- Teaching at regional training programme for clinical biochemists and for paediatric nurses.
- Instructor on DCH clinical examination course. Teaching for MRCPCH clinical exam.
- Teaching at hospital induction for senior house officers.
- Personal tutor for undergraduate medical students.
- Regular bedside teaching and problem-based teaching for medical students on the wards.

I am also actively involved with the assessment of junior colleagues and providing them with appropriate feedback.

Management and Administrative Skills

- I have received awards for the winning business case presentation on 'stepping up to your consultant role course'.
- I am experienced in organising tasks for junior medical staff on a day-to-day basis. I have organised rotas for SpR and I am currently coordinating a rota for 20 core/foundation trainees.
- I have participated in the interview process for ST posts in the HENE LETB.
- I have organised various teaching programmes including those for MRCPCH clinical examination.
- I have participated in departmental directorate meetings and diabetes team meetings.
- I have organised and chaired junior doctor's communication forum for the department of paediatrics.
- I am currently in the process of producing an annual report for the neonatal unit.
- I am responsible for organising various endocrine tests for the department.
- I have been responsible for organising my own outreach clinics during attachment in community paediatrics.
- I have experience in managing neonatal intensive care costs for the purpose of accepting retrievals and

intrauterine transport from other hospitals. I have managed antenatal folder.

Parent Information Leaflet

I have produced the following information leaflets for parents:

- Parent information booklet on NNU at Hospital as part of a group.
- Pneumococcal vaccine in children. Information for parents.

Clinical Guidelines

I have produced the following clinical guidelines for the Paediatric department:

- Guidelines for the use of the Pneumococcal vaccine in the Neonatal Unit.
- Protocol for admission of asthma directly to the paediatric ward from A&E.
- Evaluation of children with suspected immunodeficiency.
- Evaluation of the Child with Global developmental delay.

Research

I have developed, written and managed the following single centre randomised controlled trial at (name) Hospital, while continuing with my clinical duties:

"A comparison of venous versus capillary measurements of tobramycin serum concentration in children with cystic fibrosis."

Children participating in this study also complete a questionnaire indicating their preferences for either venous or capillary samples. I have obtained ethical approval for this study and so far we have recruited seven children.

Publications

List your publications in Vancouver style.

Presentations

International:

List your presentations.

Regional:

List any regional presentations.

Computing Skills

I am computer literate and regularly use Microsoft Office packages like PowerPoint, Excel and Word. I am also experienced in the use of Hiss and PACS system in hospitals.

I am competent in the use of the Internet for literature searches to enhance my knowledge.

Personal Interests

I enjoy playing tennis and skiing. I also enjoy music, watching films and travelling.

Career Intentions

I would like a job as a consultant in general paediatrics whilst pursuing a subspecialty interest in diabetes and neonatology. I would also like to maintain my skills in paediatric endocrinology where possible.

Referees

Pre interview visit

If shortlisted, a pre-interview visit is essential for success.

Take your CV when visiting and put your NHS job application number on it. The NHS job applications are anonymised. The person shortlisting would not know your name or other personal details. So remember to put your NHS job application number on your CV.

Who to visit?

- Colleagues in the same specialty
- Consultant Colleagues in an allied specialty e.g. anaesthesia, radiology, nephrology, surgery, gastroenterology etc to see what they feel about the new appointment
- Clinical Director
- Medical Director
- Chief Executive of the trust

What will you talk about?

Questions for the consultant colleagues

- What are you looking for in a new consultant?
- How do you envisage things will change?
- Are there any deficiencies in the service you want addressing?
- Are there plans for specialist nursing roles?
- How is 'X' currently provided? E.g. Are there any acute GI bleed services?
- Any plans for community clinic provision?
- Is there opportunity for research in the department?

- How does acute medicine/surgery have an impact on specialty?
- What junior support do you have?

Questions for the chief executive

- What are you looking for in a new consultant?
- Are any major service reforms being planned?
- What is the financial position of the trust?
- What is the likelihood of investment in the (specialty) service?
- Are there any key nursing/ recruitment issues?
- Are there any local political influences?
- What are relations like with commissioners?
- Are there any specific targets the trust wants to focus on?

You might ask the chief executive about the trust strategy for the next 5-10 years, what plans exist for short term developments, what the role is of the clinician in management, whether there are any threats to the service, PFI issues, Foundation Trust status etc.

Questions for the medical director

- What are you looking for in a new consultant?
- Are there any major service reforms planned?
- What is the framework for trust clinical governance?
- What is the role of the specialty in acute medicine/surgery?
- What is the role for acute physicians in GIM?
- What are the teaching opportunities?
- What are the links with the medical school?
- What are relations between clinicians and management like?

- Is there a scope for service development?

Ask the medical director about strategies for R & D, place for R & D in business plans, progress towards Foundation Trust, the quality of the relationship with commissioners, financial status etc.

Tips

- Applying for a locum job? You are within your rights to ask: why locum—will there be a substantive post here? And if so have you got someone under consideration for that substantive job?

- Don't be afraid to ask about resources, facilities, local schools etc.

- Don't try to impress. It's more about whether this is really the job that you want (unlike the interview), and if you go in there with that attitude, you'll impress more than if you go in trying to impress. This is a two way process. They need to sell the trust to you, and you need to sell yourself.

- It's worth remembering that the pre-interview visits are your opportunity to see if you really want the job - you can ask what you wish of whoever. You need to be 100% certain about the job before you go to the interview - so think of all the things you want to know and then consider who could answer the questions best.

By the end of the visit, you must know all about the department, how it works, what services it lacks, what needs improving, what are the challenges and what changes are ongoing.

Next time you meet will be at the interview, when they will ask you 'What will you bring to the department?' Your answers must be what they want to hear!

Interview Process

Who will interview you?

The make-up of an Appointments Advisory Committee (AAC) is laid down in a Statutory Instrument (The National Health Service Regulations 1996 No 701).

The core membership of the committee is as follows:

- A lay member (non-executive director)
- College assessor
- Clinical Director
- Chief executive and/or Medical Director
- At least one consultant from the employing trust
- If the job includes a substantial research or teaching commitment, a further representative of the local university may also be present.

The questioning usually follows a set pattern and in the following order:

- College Representative
- University Representative
- Hospital Consultants (usually 2)
- Medical Director
- Chief Executive
- Lay Chair

College Representative: is usually from the same specialty but from outside the region. He/she is there to see if you are appointable. They could ask about your specific training for the post for which you are applying.

University Representative: is only present if there is a major teaching commitment so you will be asked about teaching. He/she will also ask about research interests, publications, why/why not an academic post.

Hospital Consultants: You should know them already from pre-interview visits. They will ask what you will bring to the post, how you will improve their department, how you will fit in their team, etc. They are the most interested parties there.

Medical Director: is concerned with clinical governance and safety. He/she may ask you about audits, complaints, safety concerns etc.

Chief Executive: How will you benefit the trust? He/she will ask about your management experience and your understanding of NHS policies. Where do you see yourself in 10 years' time?

Lay Chairperson: Usually a non-executive member of the trust board. They will run the interview. The lay chair will want to know a bit about you outside medicine.

The interview will last around 45 minutes or more. The interviewers will have clarified their objectives in general and decided on a format and a set of questions for each interviewee. Commonly, appointment advisory committees take a formal structured approach where the interviewers take turns to ask questions reflecting their particular interests. In the final stage you will be given the opportunity to ask questions (don't – you will have asked these in the pre-interview) and the Chair will inform you of how the result will be conveyed.

Remember, there are only three true job interview questions:

- Can you do the job?- Strengths
- Will you love the job?- Motivation
- Can we tolerate working with you?- Fit

Psychometric Tests

Increasing the number of trusts employing psychometric tests as part of the consultant recruitment process.

What is the purpose of psychometric tests?

These tests (sometimes known as personality profiling) are used to assess the ability, performance and behaviour of the candidate. These tests can provide an insight into assessing personal qualities such as personality, beliefs, social competence, values and interests and measure motivation or drive.

They are used to provide employers with a reliable method of selecting the most suitable job applicants or candidates for the job advertised.

How is a psychometric test conducted?

The test generally consists of a series of multiple choice questions. Questions focus on specific traits of the individual's personality.

The questionnaire is subsequently assessed by a psychologist or other trained assessor. The questionnaire may also be followed by an interview with a psychologist/trained assessor to validate the results further.

How do psychometric test results impact on consultant recruitment?

The tests provide additional useful information to inform the final selection of the candidate. The test results may be used in the interview to guide specific questions.

These tests are not about passing or failing but about giving a profile of the candidate to the employer, whose task it is to match an applicant to the job or work place.

Can you prepare for the psychometric test?

No, because there are no right or wrong answers. Further, each department looks for specific personality traits in the candidates and trying to second guess the answers may be counter-productive.

Tip: Try and provide honest answers rather than perceived best answer lest you come across as inconsistent.

Ref: http://www.psychtesting.org.uk

Interview Tips

The emphasis of the interview is:

- What is good about you?
- What have you achieved?
- Where does your potential lie?
- How can you integrate?
- Why are you better than the other candidates?

What are your objectives?

- To gain the initiative
- To present yourself in the best possible light
- To make known your talents and expertise

What are the interviewer's objectives?

- Find the most suitable person
- Encourage you to express yourself fully
- Look for specific skills and achievements
- Sell the job and the organisation
- Assess your initial impact and social fit

Tips

- You want to convince the panel that you will bring enhanced benefits to the organisation. Candidates will be remembered if they are distinctive, have something interesting to say and can make a unique contribution. Therefore, consider what you have got that makes you special and what makes you fit in.

- Review the job description and identify how you would fulfill each of the key tasks outlined. Watch you are not seen as overconfident. Your initial impact is vital.

- When you enter the interview room, remember to close the door behind you, walk forward confidently, respond to offered handshakes, wait until you are asked to sit and remain quiet but alert for the opening question.

- Remember to dress smartly and appear well groomed.

- Use positive body language e.g. rest your hands on your lap: be comfortable and relaxed; keep your head raised and listening; make constant eye contact and smile. The overall demeanor should be "confident humility".

- Avoid negative body language: fidgeting, biting nails, crossing arms, clasping chair, getting distracted, tapping feet, etc.

- Remember to project yourself confidently in a clear, controlled and steady voice that can be easily understood. Use a range of tones, pause before speaking and speak slightly slower than normal.

When answering questions:

- Listen carefully to questions. Don't make them ask it more than once.

- Make eye contact with the interviewer before speaking. Address the person who asked the question. However, glance around to engage the whole panel.

- Adopt a relaxed posture; sit squarely in the chair.

- Keep to the point; aim to be precise; give a full answer and do not waffle. Give a framework - 'Three aspects to this; one...'

- Structure your answer; it should be logical and clearly understood.

- It is fine to say 'I don't know'.

- Avoid using jargon.

- Speak confidently so that you can be heard,

- Do not speak too quickly or slowly.

- Be enthusiastic and positive.

- Give evidence of what you say - use your track record of success - don't be too modest but don't make it up. Give plenty of work related examples.

- At the end of the interview take your leave as smoothly and politely as possible. Do not add any afterthoughts and thank the panel for their time through the chair.

Top tips

- Remember: performance and, ultimately, your success, is a result of thorough preparation. Take time to practice. Practice develops performance in all things; interviews are no exception.

- Get a colleague to give you a mock interview.

Preparing for the interview:

- Be prepared to do at least 10-20 hours of reading.

- There are hundreds of possible questions, but they all boil down to 10-15 themes (discussed later). It is vitally important that you spend some time brainstorming your ideas on each of these topics. Take four or five questions from each theme (Teaching, research, difficult colleagues etc) and brainstorm them. Then see how you can structure your answer using 3 or 4 bullet points. This would help you recall them on the day of the interview. Don't try and write down all your answers word by word. There is a real risk that you will sound rehearsed and, worse, may forget the content of your answers.

- Don't try to memorise answers.

- Remember, interviewers generally allow up to 2-3 minutes per question. So you should aim for 1.5-2 minutes for your answers allowing time for the question to be asked and supplementary questions.

Formulating your answers:

- Answers should be 1.5-2 minutes long

- A good answer generally has 3-4 keys points.

- A good answer should be your bullet points and then expansion statements. Provide your message at the beginning of the answer, if possible. You can substantiate the message in the later part of your answer. For example, 'I am a really valued member of the team because....'

- Provide objective examples to substantiate your statements. This is very important. For example, it is no good saying that you are a team player unless you provide examples to back up your statement. An appropriate answer would be: 'I am a good team player. In my last 360 degree appraisal I was rated very highly as a team player. Furthermore, my consultants often praise me for being such a great team player.'

- In every answer you give, look for the opportunity to show the panel just how much wider reading you have done.

Are interview courses any good?

- Attending a course helps inasmuch as it allows mock interviews. You can achieve the same objective if a consultant or senior colleague is willing to help you with mock interviews.

- If you were to attend an interview course- attend a course with a small group. Small groups would allow you to discuss good and bad answers and obtain personal feedback on your own technique.

- If you are attending a course, attend only after you have done the basic preparations yourself.
- Interview courses can only fine-tune your preparation. They aren't designed to organise your preparative thinking from the bottom up.

Dressing the part

- Men could wear a conventional pinstripe suit and white shirt.
- Women could wear a simple black dress (skirt/trousers).
- Hair should be clean and dry. Fingernails should be clean or painted with a pale colour.
- Glasses should be freshly polished.

Interview Presentation

You may be asked to give a presentation to the interview panel. This is common at consultant interviews. This is normally for 10 minutes or less. Prepare no more than 6-7 slides for a 10 minute presentation.

You will be informed of the topic and length of the presentation. Plan your presentation thoroughly: practice your delivery and use audio-visual aids which you feel comfortable with (and check what equipment will be available on the day), keep to time limits and anticipate questions (even suggest a few in the presentation).

Tip: take some print-outs of your presentation in case technology fails on the day!

Common topics for presentation

- Discuss the impact of current NHS changes on your specialty
- Discuss the issues affecting your specialty
- How would you set up a new service for a particular condition?
- Expensive new therapies and affordability
- How can you improve a particular service in the trust?
- Discuss what can you offer the trust

Your presentation needs to achieve the following:

- Convince the panel you are qualified and experienced
- Outline your possible contributions on a strategic and detailed level
- Establish good relationship with the panel

Be sure to inject:

- Professionalism, pace and drive
- A degree of formality
- Controlled enthusiasm

Background and CV

Tell me about yourself.

Talk us through your CV/Application form.

Tell us about your background.

These questions are essentially similar and often are the first questions asked at the interview. Use this question to emphasise your positive points. It is a good opportunity to utilise the vagueness of the question to make a good impression.

Remember, you and your CV are a lot more than just medicine. So in answering the question, dwell briefly on the various aspects of your experience including:

- Clinical (your training so far and the skills acquired)
- Academic (teaching, research and audit)
- Generic skills (communication, team player, leadership skills)
- Social (family, hobbies etc.)

Finish your answer on an enthusiastic note by spending a few seconds on your career plans.

Summarise your specialty experience over the last five years.

What has been the extent of your specialty training to date?

Focus on the relevant bits from the first answer above.

So do you think you are independent at all endoscopic procedures?

So you think 5 years is enough to acquire competence in your specialty and GIM?

Have you had experience dealing with all major medical emergencies?

Yes

- Focused and competency based training

- Learning does not stop with the end of training. What you are saying is that although you are competent and positively excited at taking the next step up to become a consultant, you do appreciate and believe in lifelong learning.

- Reflective learner. This has helped you identify the gaps and lacunae in your training/knowledge/skills, leading you to take steps to develop yourself further.

Give me a summary of your general medical training to date.

What is exceptional about your CV?

What part of your CV are you most proud of? Why?

Here is your chance to demonstrate your enthusiasm for a particular aspect of your CV. It could be

- Research experience

- Teaching experience

- Management experience

- Any other skill

If you could change one thing on your CV what would it be?

This should ideally be in line with your future plans. This would convey that you are a go-getter. So you could talk about:

- More research experience/more publications/writing your MD

- More teaching experience/formal teaching qualification

- More management experience

- More overseas experience

- Anything else

What is the most difficult aspect of your current post and why?

You need to mention something with a positive spin. Something like a week of night on call. You could buttress your point with the Royal College guidance of having 4+3 nights on call. Talk about patient safety/quality care, if you wish.

If you were given the opportunity to withdraw from GIM, would you like to do this?

This depends on the plans at the hospital where you are being interviewed. So, if your future colleagues intend to withdraw from the GIM rota, don't sound too enthusiastic about GIM.

Alternatively, you could say that you enjoy GIM but would like a better balance between GIM and your specialty

Are there any gaps in your training to date?

No.

You should elaborate by saying that being a reflective trainee has helped you in identifying what you don't know. This has ensured that you have identified gaps and taken corrective measures. Quote examples like attendance at a particular course subsequent to your reflection and identification of a training need.

If you were to start your career again what would you change?

You could spend more time on research, attain formal educational qualifications, expanded role in management or overseas experience etc.

What are your ultimate career intentions?

- To be a good doctor
- Increasing role in education/research

- Increased role in management to influence change

What are your interests outside of medicine; do they have any impact on how you practice medicine?

What are your hobbies? How do they influence your medical practice?

Relate your hobbies or sports interest to your practice of medicine. You could say that it helps relax you, keep you fit and maintain a work life balance.

I really enjoy most team sports but I don't get a lot of time to indulge myself.

Apart from team sports, endurance sports are seen as a sign of determination: swimming, running and cycling are all OK.

Games of skill (bridge, chess and the like) demonstrate analytical skills.

Would you rather work in a shift pattern or a traditional 24 hour on-call pattern?

Mention the pros and cons of each and then say you prefer a shift pattern (as that is what most hospitals have in line with EWTD).

24 hours on-call allowed for better continuity of care but were very tiring.

Shift pattern allows for more senior input and helps patient care. The downside of continuity of care can be overcome to a large extent by good handover practices.

If you were able to do one thing that could improve the wellbeing of the world/mankind, what would you do?

Education; the only way to change (and improve) the world is by changing the human being. Education helps you achieve the goal.

What information technology skills do you possess?

My information technology skills are commensurate with my requirements. I am fluent in the use of MS Office, PowerPoint, Database and the internet. I can conduct a literature search and am particularly interested in the use of Web 2.0 tools to improve education and training.

What was the most important event in your life?

You could be personal here. The question is about the content of your life. So you could mention the birth of your son/daughter, climb to the Mount Everest etc.

Why is there no separate Clinical Governance section in your CV?

Discuss clinical governance and your commitment to it. Then elaborate that your CV contains all the components of clinical governance (so a separate section was not needed). Mention audits, courses attended to improve your education and training, any guidelines you wrote (demonstrate your commitment to evidence based medicine and clinical effectiveness) etc.

Personal qualities, motivation and drive

What made you go into medicine?

I enjoy working with people; it's one of the things that attracted me to medicine. It gives me pleasure to help people. It's satisfying when people are pleasant, but I also enjoy the challenge of working with difficult people.

I have always been very strong on the sciences, but interested in people and social issues too and more and more medicine seemed like the suitable choice.

What is your career ambition?

Where do you see yourself in 5/10 years' time?

Your career ambition

- Consultant in University Hospital/DGH
- Patient care and communication – maintenance and increased skill, expansion of skills and interests etc
- Colleagues- foster good working relationships
- Others – interests/IT/research/NHS/leadership
- Teaching and training- become the Royal College tutor, lecturer in the University, a role in the LETB
- Programme director etc

Why do you want to do this specialty?

For example for gastro:

- Good combination of interesting medicine, practical procedures and one that would involve a good mix of inpatient and outpatient care.
- There is a wide choice of sub-specialties, all of which are readily encountered in every region.

- The technology in endoscopy is rapidly advancing and there are always new things on the horizon to keep you stimulated.

- There are wide ranging options for research in basic science and clinical areas and funding appears to be healthy as many of the NHS targets come under the umbrella of gastroenterology.

- Lastly, I have never met a disillusioned gastroenterologist. They appear to be very happy with their choice!

What do you like about this specialty and what do you dislike?

Again, taking the example of gastro:

Likes

- Lots of hands-on work
- Multidisciplinary working
- Choice of sub specialties

Dislikes

- Lack of GI bleed rota in many hospitals
- No NSF for GI diseases

How would you dissuade somebody from entering this specialty?

I suppose more than dissuading; I will present the strengths and the weaknesses (if any!) of the specialty and let them make a decision for themselves.

What are the challenges facing this specialty over the next ten years?

Show your forward vision as to how things may change and how you are going to adapt.

Quoting gastro as an example:

- Most gastroenterologists continue to practise with a commitment to general internal medicine. The future may see a clash between the ever-burgeoning internal medicine demands and heavy sessional commitments to gastroenterology and demand for a GI bleed rota, with the possibility that the non-GI medical load may devolve to specialists in acute general medicine.

- Despite the advent of sophisticated imaging, the place of endoscopy in gastroenterology will remain pre-eminent owing to the need for samples and for the ultimate in minimally invasive therapeutic procedures.

- The Bowel Cancer Screening Programme continues to roll out with an accelerating pace of JAG accreditation visits. These JAG visits will result in considerable and often long overdue investment in endoscopic facilities up and down the country and this is very welcome for all concerned.

How do you know you are making the right career choice?

A career to me is a way of life and not something that pays the bills. So I suppose if I am happy doing what I do, I have made the right career choice.

Why do you want to join our Trust? (Important questions-prepare well)

Why this hospital?

Why should we recruit you rather than any other candidate?

What have you done that is different to anyone else?

What makes you a good candidate for the job?

This is where all the research from trust websites, care quality commission websites and the information you gathered at your pre-interview visit will be useful. Just tell them what they want to hear. Marry up your skills to their requirements.

I think I am the ideal person for the job because I believe I have the skill and experience to do the job reliably and efficiently. Next, outline your key strengths in education, research, management or any particular skill in your specialty.

Apart from the hospital and how great it is, you can say that you have looked around the area and like what you see. Try and give examples of good points even if it is good connections with major centres or good schools etc.

What would you want from the hospital?

Tell them about your special skills and how you plan to use them to develop the department and the hospital further.

What do you have to offer us?

Tell them what makes you special. Explain your education/research/management or any other special skills. Again your answer will be guided by your research prior to the interview. For example, if the trust was planning on increasing the intake of medical students; your educational qualifications and experience will be highly desirable. You will need to highlight this skill.

Give us three adjectives that describe you best. If I asked the people who know you well to describe you, what three words would they use?

If I were to phone your junior staff and ask what you are really like what would they say?

What would your friends say about you?

- Easy to get on with and quite outgoing
- Good sense of humour and generally cheerful
- Committed and enthusiastic
- Good team player

What kind of feedback would I obtain from your patients if I asked them?

Compassionate, patient and competent

A good doctor!

What would you like written in your obituary?

When he came, he cried and the world rejoiced.

When he went, he rejoiced and the world cried.

Describe yourself in as few sentences as possible?

What are your main strengths?

Give 3-4 strengths appropriate for the job:

- Good communicator
- Good interpersonal skills
- Enthusiasm
- Keeping calm under pressure

I believe I have demonstrated this in the past—give examples.

What is your main weakness?

Admit a minor weakness. But present your weakness with a positive spin. Practice responding to this question. It'll take a few dry runs before you sound succinct and articulate. Options:

- My family would probably accuse me of being a workaholic because I can't relax while there's something that needs doing.
- Tendency to take your work home (which you can resolve with the help of your family and personal will).
- Impatience- I know I could improve my patience when working with people who don't work at the same pace as I do. What I have found is that by helping such members, I

can move the project forward instead of being frustrated and doing nothing.

- My biggest weakness? I would say chocolate, especially milk chocolate.

- No research background- but you can still say I can critique a research paper and teach my juniors to do the same.

What skills have you gained that will make you a good doctor/consultant?

What are the qualities of a good consultant?

Qualities of a good consultant:

- He should be a good doctor. He should be competent with good teaching and training skills. He should endeavor to support research and audit.

- He should have some leadership qualities and good management skills.

- He should be flexible. We work in a complex environment. Flexibility is needed to balance best practice and limited resources and the interface between two independent services

- Sense of humour

- It is not possible to have all the above qualities, so consultants need to strive for continuous development and improvement

What is the difference between a SpR and an SHO? You've done a locum appointment in a training post—what was the most important step up you had to deal with compared to being a senior house officer?

As a senior person, you need to have more of the soft skills:

- Managing a team

- Delegation
- Negotiating skills
- Leadership

Name two skills that you would like to improve over the next two years.

What skills do you need to develop most?

- Emotional intelligence (E.I) - E.I. is the capacity to read others by just observing non-verbal behaviour and to be able to act appropriately on the information to the benefit of the person and the service.

- Teaching/research/managerial skills

What job have you particularly liked/disliked?

As a trainee I have tried to derive most from my jobs. Yes there were jobs (if any!) which were not ideal but on discussions with the local supervisor, we arrived on a mutually acceptable solution. Give an example. You need to convey an impression that you are a problem solver!

How do you measure success?

Success at work- my main measure of success is feedback from patients and colleagues.

Would you be happy being an average consultant?

I have always strived for excellence. If you look at my CV, I have been an active participant in improving things (quote your audits, any guidelines you have produced). Similarly as a consultant I would strive to improve the department by improving the services where I can and bringing in new services to provide world class care. Always mention your enthusiasm for teaching and training and your belief that no service can operate successfully without the integration of teaching and research in daily activities of service providing.

Would you like to become a clinical director?

Yes. I believe the clinical director plays a vital role in performance management of other consultant colleagues as well as providing a vision for the department. I find this very exciting. I believe I have a lot to contribute and would like to consider the role actively in the future.

Having trained under the Calman (or MMC) system, do you feel prepared?

Absolutely! I think 5 years is a reasonable time for training as long as you are focused. I have been a reflective learner and this has helped me identify the gaps in my knowledge or skills. Besides, I believe that learning is a lifelong process of education and continuous improvement. While I feel confident that I am ready for independent practice, I do appreciate that learning does not stop here

What concerns you about this job?

What do you think will be your biggest challenge in this post?

Mention something (like developing a new service or any particular problem the department may be facing) which would be difficult. Say in the same breath, however, that you are positive that with enthusiasm, persistence and teamwork, the job can be done.

What are you hoping to gain from this post?

I am looking for the opportunity to accomplish my best work. So I suppose I would say that what I am looking for is a progressive trust that will provide a challenging, stimulating and supportive environment for its employees and their achievements.

How would your consultant/seniors motivate you?

Tell us about your best consultant/colleague.

A caring and compassionate doctor who believes in teaching and training and with a positive attitude; mention a consultant who was compassionate, competent and approachable besides being a good teacher.

Tell us about your worst consultant/colleague.

When I say worst, I mean I did not like certain qualities. Mention something like not being approachable or a poor trainer.

What are your biggest accomplishments?

Although I feel my biggest achievements are still ahead of me, I am proud of my involvement in (quote any example). I made my contribution as part of that team and learnt a lot in the process. We did it with hard work and dedication.

What experiences outside medicine have you found useful in your medical career?

Mention something you are proud of...like leading a team of trainee doctors to a disaster-hit area. You could say that it developed your interpersonal skills. You also realized that appealing to the good side of people always brings out the best.

What sort of hospital would you rather work in and why?

Teaching hospitals as you have more opportunities to develop

DGH- if dealing with patients/people is your forte.

How do you relax?

There are numerous ways in which to relax, so try and be a bit more specific. If you play a sport, for example, tell them who you play for or how often.

How would you balance extra-curricular activities with being a consultant?

Good question to discuss work life balance. Discuss the importance of relaxation and de-stressing etc.

Rate yourself on a scale of 1 to 10?

Say 8 or 9, saying that you always give your best, but that, in doing so, you can always increase your skills and therefore always see room for improvement.

What would you change about yourself if you could?

I would find it easier if I could get down to the gym three evenings a week. I always wished I'd learnt Spanish/violin/to play tennis properly. I still intend to get around to it one day.

What makes a good leader?

I believe a good leader is a matter of motivation. Good leaders are people who can keep a team enthusiastic and committed to success despite difficult and challenging conditions. However, I have found that to be a good manager, it pays to be as versatile as possible, depending on the situation.

What skills are required for good leadership? What leadership skills have you acquired during your training?

- Effective communication and interpersonal skills
- Competence
- Leading by example
- Nurturing the team- social/pleasant/approachable/easy to talk to/receptive to new ideas
- Encouraging others to recognise each individual's contributions to the success of a project
- Keeping cool under pressure

What does leading by example mean to you?

It means to behave in ways that are consistent with professed values and help others to achieve small gains that keep them motivated, especially when a goal will not be achieved quickly.

Can you say 'no'?

Yes, if I am being asked to do something that is beyond my remit or in the time frame that is expected. It is better than agreeing to do something and then finding that you can't deliver.

My colleagues in the past will acknowledge that I only say no when I really do have a good reason and not just to avoid work or to be unhelpful.

What is the difference between a manager and a leader?

I suppose I see managing more as organizing work and controlling resources, whereas to me, leadership is more about coming up with new ideas and, above all, being able to take people with you, whether this is motivating people to tackle something new or keeping morale and communication going through troubled times. I think it is important to see things through, as well as having innovative ideas.

The difference between leadership and management can be illustrated by considering what happens when you have one without the other.

Leadership without management

...sets a direction or vision that others follow, without considering too much how the new direction is going to be achieved.

Management without leadership

...controls resources to maintain the status quo or ensure things happen according to already-established plans. E.g. a referee manages a sports game, but does not usually provide "leadership".

How would you motivate others?

I think good motivational skills are essential. Getting the team to work towards a common goal with enthusiasm and purpose is vital for performance.

I make sure that my team has clear goals and targets, and understand how those contribute to the overall aims of the trust. They know why their role is important and how it fits in with the rest of the organisation. All the members are kept informed about developments and, where appropriate, involved in discussions and contribute to the decision making process. I also make an effort to understand the personal motivations of my team members. As a result, my juniors are, I believe, well-motivated and work well both individually and as a team.

Do you consider yourself a natural leader or a born follower?

I would be reluctant to regard anyone as a natural leader. I believe leadership is not just a set of exceptional skills and attributes possessed by only a few very special people. Rather, leadership is a process and a set of skills that can be learned. Leadership requires, first of all, the desire; then it is a lifetime learning process.

What is the difference between leadership and management?

Management means:

- Getting things done with and through others
- Achieving goals through efficient use of resources
- Achieving objectives within a series of constraints on time, money and other resources, using a range of techniques and processes

Leadership is:

- The ability to envisage a new path forward and inspire others to make the vision a reality

- The skill of influencing and motivating others to work together to achieve goals

- The art of remaining true to values, ideals and achieving goals, overcoming hurdles and constraints, no matter how difficult that may be.

Leaders do the right things and the manager does things in a right way.

Communication and Team Playing

How would you rate your communication skills?

Give us an example of a situation where your communication skills made a difference to the care of a patient.

Give us an example of a situation where you failed to communicate appropriately. Tell me about a situation where your communication skills did not succeed in getting something done.

What skills have you acquired that make you a good communicator?

How can you improve your communication skills as a leader?

These questions have the same theme.

Rate your communication skills as good or effective. You have to back up your answer with some specific examples where your communication skills made a difference. Try and make it specific to the specialty you are applying for: for example, in gastroenterology, you may wish to focus on relationships with other colleagues like surgeons. You may also discuss breaking bad news, rapport with patients etc.

Skills acquired- 'I have acquired good listening skills. This has helped me to build rapport with my patients, helped me understand them better and overall provide improved care.' You may also wish to discuss other skills like negotiation skills, ability to explain complex details in simple terms.

You can always improve your communication skills. You can improve them by either observing your senior colleagues or attending the relevant training courses

What are the attributes of a good team player?

Good team members are communicative, supportive of the other members, flexible (they can fit in with others and adapt to changing demands), unselfish (they put the needs of the other team members on a level with their own) and are interested in the success of the team as a whole, not just their own performance

Are you a team player?

More and more we work in teams so the answer should be 'Yes' but give examples as to how you know, e.g. feedback from your consultant at appraisal, 360 degree appraisal.

Do you work better as part of a team or on your own?

I would say from past experience that I enjoy being part of a team. I like the camaraderie and that feeling of everyone working together towards a common goal. I believe a good team member should be competent, communicative, supportive, flexible and unselfish and I try to demonstrate these qualities when working with others. Give an example where you worked well in a team, demonstrating at least one key attribute of a team player. 'However I am quite happy working alone when necessary- I don't need constant advice and reassurance but I prefer to work in a team as so much more is achieved when people pull together.'

Give us a recent example of a time where you worked as a member of an MDT.

You could say something like: 'More and more we are dealing with elderly patients with complex care needs. I liaise with the OT/physio/family/nurses/social care workers to ensure a thorough management plan is in place to ensure optimum care.'

What makes a good team?

I believe a good team is one where the members are committed to each other and to the successful achievement of a goal. I believe it's my role as the team leader to ensure that the team is more than just a collection of people working on the same ward/hospital but a real team pulling together with a strong sense of cohesion. I ensure this in my team by ensuring that everyone knows what their role is within the team and how important it is to the overall outcome. I make sure that individual skills and input are valued and appreciated not just by me but by everyone concerned. I encourage team members to support each other to complete tasks rather than focusing on their own responsibilities. I also support team unity by organising social and bonding activities.

A number of studies have revealed the following characteristics of successful teams:

- A meaningful, clearly-defined task
- Clear team objectives and individual targets
- Regular meetings
- Regular feedback to individuals and the team's success in achieving objectives
- The right balance of people
- Reflexivity – the ability to reflect on team performance and adapt and change
- The experience of full participation, which reduces stress and may lead to better care
- Good leadership

A good team shows excellent internal and external communication.

External communication means keeping the team in touch with what is happening in the wider organisation, and letting the organisation know about the team - how it's meeting its goals, what resources it needs, how it is innovating, etc. It is maintaining these external links which keeps the team in line with the organisational goals and which lets the rest of the organisation know about the role of the team. This will usually be an important part of the leader's role.

Internal communication is naturally less formal and more constant than external. It involves making sure everyone has a voice - even the silent members of the team - so that risks can be appreciated, problems aired, and the best care given. Sometimes this will involve including the patient or the carer in the team as an honorary member so that they are given the same chance to contribute to any decisions made.

Tell us about your experience of managing a team of people.

As registrars, most of us look after a team of junior doctors on the wards and when on call. You could mention you ensured that they were well supported, understood their roles, ensured that their study and annual leave were sorted, ensuring their training needs, providing feedback as appropriate, organising team meetings etc.

Describe a situation where you had to give negative feedback to somebody.

Quote any situation but remember the general principles of giving feedback. Feedback is part of the learning cycle and is thus an integral part of any learning experience.

- Feedback should be descriptive rather than judgemental or evaluative. (E.g. this is what I saw- what do you think?)
- Make feedback specific rather than general.
- Focus feedback on behaviour rather than personality.
- Focus feedback on sharing information rather than giving advice.
- Give feedback about something that can be changed.

How to give feedback;

- The person should know that he is receiving feedback.
- Collect relevant data from others.
- Make notes prior to the meeting.

- Reinforce good practice with specific examples (so give positive feedback first. The trainee is likely to be more receptive of the negative feedback.)

- Identify, analyse and explore potential solutions for any deficits in practice.

- Encourage the trainee to self-assess their performance prior to giving feedback.

NB- If a trainee rejects any negative feedback given, explore with them why they feel like this. They may be getting contradictory feedback from elsewhere which will prevent them from acting on your feedback and suggestions.

Tell us about a situation where you had to bring a difficult person on board.

The basic problem here is generally lack of communication. So mention any situation and explain how you used your good communication skills to bring the person on board.

Tell us about a situation where you showed leadership.

A good leader

- Leads by example

- Takes initiative

- Communicates well with the team

- Has clear objectives and ensures that the team understands them

- Develops team members by encouraging active participation, valuing opinions and giving feedback

Quote any example where you showed any of the above qualities. You can quote the example: You noticed a less than perfect clinical care- decided to audit it- presented it- and subsequently helped bring the change to improve the quality of care. You can use any of the audits you did or say you initiated a teaching programme for juniors to help better train them to avoid a repeat of the bad practice.

Tell us about a situation where you showed initiative.

- You could quote you noticed a less than perfect clinical care, decided to audit it, presented it, and subsequently helped bring the change to improve the quality of care. You can use any of the audits you did.

- Or you could say that you took charge of organising holidays for junior doctors on the ward after noticing poor organisation of junior doctors' holidays.

Tell me about a time when you had to use your spoken communication skills in order to get a point across that was important to you?

Can you tell me about a job experience in which you had to speak up in order to be sure that other people knew what you thought or felt?

You can say something like you noticed a lack of educational programmes or MRCP teaching at your hospital. You ensured that you raised your concerns with the appropriate people and devised a teaching rota involving all the consultants and senior registrars.

Give me an example of a time when you felt you were able to motivate your colleagues?

You can quote any of the leadership or team playing qualities or teaching qualities to illustrate the point. I regularly teach juniors and can tell from the feedback I have received that they really feel motivated to improve them and provide safe and effective care to patients. You can quote a specific teaching episode.

What do you do when one of your junior is performing badly, just not getting the job done? Give an example?

This is basically an issue of performance gap. The issue is not to ignore it and deal with it promptly. The principles of giving feedback apply here. Steps to deal with it:

- Ensure that the person understands his role and what is expected of him
- Tell him what is expected of him
- Explain the difference between his current level of performance and the expected level of performance (performance gap)
- Agree as to how the gap can be bridged in a timely manner.

Describe a situation in which you felt it necessary to be very attentive to your environment?

All exams, especially the OSCE kind!

Give me an example of an important goal which you have set in the past and tell me about your success in reaching it?

Quote any example either medical or non-medical. It could be learning swimming or setting up a website. Go through the steps of planning, hard work and dedication.

What makes you angry?

I suppose like most people I get angry about inequalities in the world, poverty, cruelty to animals, but I don't find everyday irritations affect me. I have learnt that dealing with other people calmly and politely is less stressful for me as well as for them so that now its second nature.

Do you ever lose your temper?

I can't remember the last time I actually lost my temper. Everyday irritations don't affect me that much; there's always something you have to deal with. However, I take patient safety very seriously. For example, quote an example like poor handovers and how you spoke to all the juniors to ensure proper handovers.

Have you ever been in a situation where you have had a conflict with a colleague?

Yes, give a situation. (Example: your colleague in SpR not pulling his weight and not doing enough) In the end, I took the initiative and persuaded him that we should have a talk and try to work out an effective strategy for us to work together- we really needed to with us working together. We had a pretty frank discussion and, although I can't say we ended up the best of friends, we did gain more respect for one another's roles and we certainly worked more productively together. I was really glad I had taken the initiative.

How have you benefited from your disappointment?

Disappointments are a learning experience for me. I look at what happened, why it happened and how I would do things differently, if things were to happen again in future. That way, I put disappointment behind me and am ready with renewed vigour and understanding to face the new day's problems.

Give an example of a situation where your work was criticised?

A good way is to give an anecdote from your early career that shows you accepting suggestions calmly and reasonably and learning from them.

How would you cope with criticism or a complaint being made against you?

I don't take it personally if that's what you mean. I believe I am mature enough to handle constructive criticism. In fact, it's essential if I want to continue to improve my performance. I remember earlier in my career... (describe the event and how it arose and who did the criticizing.) I listened to what they said and I could see that they had a point. Describe the lesson you learnt and how it was useful.

One of your core trainees says he is getting bored in his job. How do you respond?

Find out what else is happening

What is demotivating him?

Discuss things and suggest remedies or direct him to appropriate resources

Stimulate/challenge him- take him to see referrals, endoscopy sessions etc

Do you like change?

Look at my CV, I came from Sudan, 3 house moves in 5 years. Change does not scare me.

If you could teach a medical student only one thing to make them a better doctor, what would it be and why?

Compassion and empathy; I believe this makes more difference in patient care than virtually anything else.

How would you handle a situation where you had a disagreement with a nurse over the management of a patient?

What would you do if a patient disagreed with your treatment approach?

I would try and understand why he/she disagreed with my opinion and try and reach a mutually acceptable decision while ensuring that the patient is not put at risk.

What kinds of decisions are most difficult for you?

It's not that I have difficulty making decisions but some just require more consideration than others.

A small example might be holiday time. Everyone is entitled to it, however careful planning ensures that the team has enough doctors at all times. I think very carefully at the beginning of my jobs when I'd like to take my holidays and then think of alternative dates. I discuss this with my colleagues and the consultant and tell them what I hope to do and see if there is any conflict. I wouldn't want to be on holiday when the team was depleted. So by carefully considering things far enough in advance, I don't procrastinate and make sure my plans fit in with my team.

How do you handle stress?

How do you normally cope with pressure?

I have always been good with stress. I was always the one who stayed calm during exams. I believe it's because I am good at planning and prioritising. I (describe some of the things you do to organize your workload- or anything you do to manage your time). For example, on-calls are generally very busy. Often you have to deal with multiple emergencies. This could be very stressful. However, I take a step back and look at the whole situation and involve my whole team. So I work closely with charge nurses to triage patients effectively, delegate simpler tasks to junior members of the team and keeping an eye on progress regularly.

[The most commonly accepted definition of stress (mainly attributed to Richard S Lazarus) is that stress is a condition or feeling experienced when a person perceives that "demands exceed the personal and social resources the individual is able to mobilize." In short, it's what we feel when we think we've lost control of events].

How do you recognise when you are stressed?

My wife tells me!

Tiredness, irritability or forgetfulness, inefficiency (taking too long to do simple tasks) etc. are general markers of stress and you can quote any alone or in a combination.

How do you resolve stress?

Delegating/sharing the workload, planning and prioritising work, asking for help and anything that relaxes you (hobbies, holidays etc).

How would you deal with a 10% cut in your budget?

By increasing productivity and improving efficiency! You will have to quote examples like seeing an extra patient in the clinic, vetting endoscopy requests to screen inappropriate requests, organising professional leave more efficiently to ensure minimal disruption of clinical work etc. Basically, you will have to involve the whole team in the effort to ensure patient care is not compromised.

How did your boss get the best out of you?

My last boss got superior effort and performance by treating me like a human being and giving me the same personal respect with which s/he liked for her/himself.

Are you ruthless?

I can be ruthless when required. I am ruthless with time but gracious with people.

I have got a huge amount of self-belief and I do believe that I am good.

How do you cope if project/things go wrong?

Firstly, I assess the steps we need to take to optimise damage limitation and deal with the immediate problem. Once the crisis is over, I analyse what went wrong and I include colleagues in this discussion so that we can pinpoint how we might avoid the same thing happening again.

Have you done the best work you are capable of doing?

I am proud of my professional achievements to date, especially (give an example). However, I believe the best is yet to come. I am always motivated to give of my best.

What are your views on health and safety in your job?

The interviewer needs to know that you are:

- Aware of the importance of health and safety
- Know about health and safety issues relevant to your job.
- Understands and follows regulations
- Has any health and safety training

Have you ever had to bend health and safety rules to get a job done?

I have never found it necessary to bend the rules, and I wouldn't expect to be asked to.

Would you say you are confident?

Yes, I would say that I am a confident person. I do everything I can, though, to support my natural sense of confidence; keeping myself up-to-date, preparing carefully for presentations/meetings etc. I've always been outgoing and self-assured, and my confidence in dealing with the people has developed naturally with experience and watching senior colleagues at work.

What difficult decisions have you made in a clinical setting?

Discuss any difficult decisions you had to make in managing a patient and explain why it was difficult. It could be an ethical issue or a lack of resources. It may be worthwhile mentioning how you would deal with it in the future or anything you did to remedy the situation.

How do you go about making important/difficult decisions?

There are 4 steps I follow to ensure that I choose the best possible options. Firstly, I get together all the facts I can. Secondly, I talk to the people involved in the matter and get their input as well. Thirdly, I examine all aspects and try to predict the possible outcomes. Lastly, I try to foresee any contingencies that might affect my decision along with any problems that might arise from it. When I have all the information, in my experience a clear option usually stand out. Taking into consideration factors such as timing, budget and so forth, it's usually possible to make an appropriate decision. For example (quote an example).

Explain what you understand by an equal opportunities policy.

My understanding is that every employee and everyone with whom we deal will be treated fairly, regardless of age, gender, sexual orientation, ethnic and cultural background, disability or social background.

You don't get on with a colleague - how do you deal with it?

- Try hard to get on with all colleagues
- Came across one, did not get on well, basic problem was lack of communication
- Uneasy about situation
- Working in same department
- Went to see him, cleared up the misunderstanding and worked together afterwards

What is your approach to resolving conflict?

Conflict is an inevitable part of all relationships and is not in itself an indicator of a poor team. It causes a team to function badly only when it is not addressed.

The conflict needs to be addressed at once. Often when the problem is pointed out early, and discussed, resolution follows. If this is not possible, then dispassionate fact-finding may be necessary.

In resolving conflict, be tough on problems, not on people. Here are some possible conflicts and suggested solutions for each. If these issues occurred in your team, what do you think the response could be?

- Is one member being blamed for all the team's problems? This is often the sign of a dysfunctional team, so look at what else is going wrong before you act.

- Is one individual behaving badly? Talk to them personally.

- Is team leadership at fault? Think about what you could do differently and let the team know how things might change.

- Has the team become divided? Reestablish the overriding team goal and bring everyone in on how to achieve it best.

- Can the conflict be used constructively? Take the problem seriously and explore all possibilities.

Think of an example of a conflict in your own team. What was your role, if any, in keeping the conflict going and in ending it? How could you best have contributed to its resolution?

Think of a conflict. Here is a step by step guide to resolving conflict:

- Identify the problem. Make sure everyone involved knows exactly what the issue is and talk it out until everyone understands what the key issues are.

- Allow everyone an opportunity to express their opinion/perspective. Ensure all participants feel safe and supported.

- Identify the ideal end result from each person's point of view. It might surprise everyone to discover that their visions are not so far apart after all.

- Figure out what can realistically be done to resolve conflict.

- Find an area of compromise. Is there some part of the issue on which everyone agrees? If not, try to identify long-term goals that mean something to everyone, and start from there.

How can you minimise/prevent conflict?

- Bring issues out in the open before they become problems.

- Have a process for resolving conflicts — bring up the subject at a meeting, and get an agreement on what people should do in cases of differing viewpoints.

- Make sure everyone understands their roles.

- Regular feedback.

Research and Audit

Tell me about your research experience.

Tell them about the research experience you have covering the following areas:

- Type of research you have done and your involvement in it. Whether you were responsible for ethics committee approval and R and D approval. Who initiated the idea and who wrote the protocol, patient information leaflet etc.? Whether you were the principal investigator.

- Tell them briefly about the research- type of research (RCT, multicentre study etc), methods and the outcomes. Tell them if you have published it and that you are the first author (if you are!)

- Explain what you gained out of research.

Is a higher degree important in modern medicine?

Do you think all SpRs should do research?

Same answer to both questions

The government has made a strong commitment to research. Research is (thus) one of the key components of clinical governance. However, it does not mean that all doctors should be involved in research. Research is expensive. Research for the sake of one's CV will only lead to a waste of resources, which can be more appropriately utilised involving people interested in research.

You can always involve yourself in research without undertaking a higher degree. You can always support research by helping recruit patients to clinical trials from your practice.

However, it is important for all physicians to understand the principles of research as it underpins evidence-based medicine. So for example, it is important to be able to critically appraise a research paper and form a judgement as to whether the conclusions presented can be implemented in your own practice.

These skills are best acquired through exposure to a degree of research but can be gained through journal clubs too or attending appropriate courses.

Tell me about your research. Assume that you are talking to a group of charity workers from your funding organisation.

Essentially tell them about your research in lay terms – the aims, the methods and how it may benefit the intended group. Try and relate it to the aims of the charity (if provided).

How much of your research is your own design and how much was guided by your supervisor?

How did you organise your research project? Did your supervisor write your grant application?

Same answer to both questions. Be honest. Tell them about the extent of your involvement in the research project. You may have just collected the data or consented the patients for recruitment. Whatever may have been your involvement, it is still a plus point that you were involved in a research project.

I am worried about your lack of research experience.

Tell them you do understand the importance of research. Quote whatever research experience you have (even if minimal). Tell them you are interested in clinical work and thus may not have produced the best research.

Go on to say that you do understand the principles behind research and research governance and have acquired valuable research skills like critically appraising a research paper by attending training sessions and journal clubs.

Why is research important?

Research is important for the advancement of medical sciences.

When would you say research is good research?

Clearly not all research is well-conducted. The quality of the research depends on

- Hierarchy of study design in the following order: RCT, non-randomised controlled trial, observational studies, case reports, expert opinion

- Study quality: assessing the study for randomisation process, blinding of the subject/investigator, use of appropriate statistics, whether the study was adequately powered, whether the study and control groups were well matched, whether all the study subjects were accounted for in the results etc.

- Generalisability: whether you can implement the conclusions of your study to your practice (this will depend whether the study group is in any way similar to your practice patients)

If you are asked to read one section in an article, which section will you read? Why?

The methods section is the most important to me because, unless appropriate methods are used to answer the relevant research question, the results and conclusions may be flawed. Of particular importance in methods are:

- Process of randomisation

- Whether the investigator and subjects were blinded

- Process of recruitment

- Whether ethics and R and D approval were sought

- Power of the study

- Statistics used and its appropriateness

- Baseline characteristics of the study and control group and whether they are adequately matched

How will you statistically make up for the unaccounted people in a clinical trial?

By using Intention to treat statistics

Basically, by using this statistic, you include all the patients recruited (disregarding any dropouts) in calculating the results.

How do you go about setting up a research project?

Steps

- Asking the right research question. Decide what specific area you wish to research. Once you have decided on a topic, do a literature review of the topic, focusing on research in the last 2-3 years.

- Finding a person/expert interested in the research question. Alternatively, you may wish to talk to a unit/person with an established research track record and check whether they have any interesting projects for you.

- Arranging funding.

- Further steps like Ethics and R and D approval.

- You can read more about research on the trainees in gastroenterology website www.tig.org.uk

When doing research, what is the one most important factor to get right?

Planning.

What you want to research and where. Planning a good project (i.e. a project that is practical, feasible and answers an important question) needs thorough planning. A pilot run is also important to ensure that everything is working according to plan.

Would you like to participate in research if you were appointed?

- You could say that your prior research experiences have kindled a strong interest and you wish to build on your experience by getting involved in further projects.

- Alternatively, you could say you do not see yourself as the principal or chief investigator but you understand the role of research and will help in consenting and recruiting patients for research.

Should all research be carried in tertiary centers or do DGHs have a role?

The majority of doctor's work in DGH and the majority of patients are treated in the DGH. So DGH are vital in any research conducted within the NHS. Besides, research is a vital part of clinical governance framework and, as such, involves all NHS hospitals

What are the current research developments in your field of interest?

Topical question

Mention one or two research developments.

What is Evidence Based Medicine (EBM)?

EBM essentially means the application of the best evidence from the best medical research to treat patients.

David Sackett et al define EBM as the 'integration of best research evidence with clinical expertise and patient values'.

E.g. of EBM- use of thrombolytic therapy in thrombotic strokes.

Pros and cons of Evidence Based Practice

Pros

EBM is essential for the application of best research evidence for the benefit of the patients.

Cons

- Publication bias: Negative studies may not always be published as journals tend to publish new and positive studies.

- 'Cookbook' medicine: Guidelines and recommendations may be too prescriptive and thus suppress clinical freedom.

- It is impossible to practice EBM with every clinical decision as there may be no good evidence to support clinical judgement.

What is your understanding of the term 'research governance'?

Research governance is about the ethos of best practice and ethical methods when researching. Research governance is about ensuring the highest standards of quality in clinical research. This covers scientific quality and standards of ethics, and all related management aspects in the setting up, conduct, reporting and progression to healthcare improvements. It makes certain that the interests of research participants come first. The DoH has published 'The Research Governance Framework 2005'. This includes for example:

- Ethics and R&D approval is a must before any research begins

- Researchers bear day to day responsibility for the conduct of research adhering to the principles of good practice

- Appropriate management of financial and other resources by the researchers

Research Governance is needed to:

- Safeguard participants in research

- Protect researchers/investigators (by providing a clear framework to work within)

- Enhance ethical and scientific quality

- Minimise risk

- Monitor practice and performance
- Promote good practice and ensure lessons are learned

What are the different levels of evidence available?

Grading of evidence

- Ia: systematic review or meta-analysis of randomised controlled trials
- Ib: at least one randomised controlled trial
- IIa: at least one well-designed controlled study without randomisation
- IIb: at least one well-designed quasi-experimental study, such as a cohort study
- III: well-designed non-experimental descriptive studies, such as comparative studies, correlation studies, case–control studies and case series
- IV: expert committee reports, opinions and/or clinical experience of respected authorities

Grading of recommendations

- A: based on hierarchy I evidence
- B: based on hierarchy II evidence or extrapolated from hierarchy I evidence
- C: based on hierarchy III evidence or extrapolated from hierarchy I or II evidence
- D: directly based on hierarchy IV evidence or extrapolated from hierarchy I, II or III evidence

Do you think Evidence Based Medicine is applicable to all specialties?

Yes

What is a systematic review and meta-analyses?

A systematic review is a literature review focused on a single question that tries to identify, appraise, select and synthesise all high-quality research evidence relevant to that question. It underpins EBM. The advantages of systematic review are:

- Large amounts of information can be assimilated quickly by healthcare providers, researchers, and policymakers
- Conclusions are more reliable and accurate because of the methods used
- Results of different studies can be formally compared to establish generalisability of findings and consistency (lack of heterogeneity) of results

A meta-analysis is a mathematical synthesis of the results of two or more primary studies that addressed the same hypothesis in the same way. It can increase the precision of a result. Meta-analysis is essentially a quantitative systematic review.

Ref- Greenhalgh T. How to read a paper: Papers that summarise other papers (systematic reviews and meta-analyses). BMJ 1997; 315:672-675

What is an audit?

Audit is a vital component of clinical governance. Clinical audit is principally the measurement of practice against agreed standards and implementing change to ensure that all patients receive care to the same standard. Audits help identify and promote good practice and can lead to improvements in service delivery and patient outcomes. It also helps ensure efficiency by ensuring better use of resources.

Tell me about your audit experience?

Discuss your audits (what audit did you do, why you did that particular audit, what deficiencies did the audit reveal, what change was implemented, did you complete the audit cycle etc.)

What is the difference between audit & research?

Audit	Research
Aims to review current practice against best practice and to implement change to improve current practice.	Aims to derive new knowledge which is potentially generalisable or transferable.
Will never involve a completely new treatment or practice.	May involve a completely new treatment or practice.
Addresses clearly defined audit questions using a robust methodology, usually asking whether a specific standard has been met. Results are specific and local.	Addresses clearly defined questions / hypotheses using systematic and rigorous processes. Designed so that it can be replicated and results can be generalised to other groups.
Generates evidence to demonstrate level of compliance with agreed standards. This may lead to changes in practice.	Generates evidence to refute, support or develop a hypothesis. May lead to development of new services or practices.

Ref- http://www.ekclinicalauditservice.nhs.uk

Tell me about the audit cycle.

Clinical audit is not a one-off exercise- it is a continuous cycle of quality improvement.

An audit cycle or spiral is the process of setting standards, systematically evaluating the care of service with respect to the standards, changing practices to improve the care or service and then reevaluating the resultant care or service. In some cases, the standard will have changed between the audit and re-audit. Hence, an audit cycle is sometimes referred to as audit spiral.

What problems are there with the way audit projects are carried out?

- Very often not done properly: audit questions are not clearly defined or methods are not robust
- Not completed
- Audit cycle not completed
- Audits not presented
- Audits done for the sake of a CV, hence do not lead to change or improved services

Have you heard about the quality improvement (QI) project?

As mentioned above, most audit projects achieve little. The trainees learn little and make no difference to their own practice or the experience of their patients. They learn little about the real power of quality improvement in practice.

Learning To Make a Difference (LTMD) is an initiative to enhance the training of core medical trainees to enable them to learn, develop and embed new skills in quality improvement and put these new skills into practice.

A QI project in a nutshell involves identifying a clear and focused aim, deciding on change(s) you are going to make and deciding what you are going to measure before you start to monitor the impact of any change.

The expectation from August 2012 is for all CMT trainees to do a quality improvement project.

Ref: http://tinyurl.com/c53jbyq

What research project would you develop first in the post?

This will depend on your research interests and the research set up at the trust you are applying for a job.

What area of research is important for the future?

You could relate an area in your specialty or alternatively you could mention research in killer diseases like malaria (need for vaccine), improved sanitation, air pollution etc.

Teaching and Training

What is e-learning?

E-learning comprises all forms of electronically supported learning and teaching. It is essentially a computer and network-enabled transfer of skills and knowledge. E-learning applications and processes include web-based learning, computer-based learning, virtual education opportunities and digital collaboration. The content is delivered via the Internet, intranet/extranet, audio or video tape, satellite TV, and CD-ROM. It can be self-paced or instructor-led and includes media in the form of text, image, animation, streaming video and audio.

The idea is that learning is not based on objects and contents that are stored, as though in a library. Rather, the idea is that learning is like a utility, like water or electricity, that flows in a network or grip that we tap into when we want.

Why is e-learning needed?

- Shorter working hours for the trainee

- Increasing complexity and range of medical conditions

- Increasing patient awareness

- Insufficient time to develop an apprentice style relationship with trainers because of short duration of attachments

- Financial pressures to make training cost-efficient

- Need to improve standards in some areas

- Standardise training and spread best practice

What are the advantages of e-learning?

- More patient-friendly (vs. Apprentice model)

- High quality (vs. "see one, do one, teach one")

- Efficient (reduced error and training time) and effective (reduces learning curve)

- Personalised learning: learner-centric and owned by the learner

- Interactive learning in a network

- Not constrained by geographical location of trainee and trainer

- Training by committed experts accessible to all

- Mobile learning

Tell us about your teaching experience?

Be descriptive. Show your enthusiasm for teaching

- Discuss who you have taught and how often you teach and whether it is formal or informal teaching. Discuss some of the teaching methods you use. You can quote some interesting topics or episodes

- Mention any teaching courses or qualifications

- Discuss how you know you are a good teacher. You could say that you regularly arrange feedback from your students or occasionally get your teaching session peer-reviewed. You could also mention that your teachings skills were specially commended in your 360 degree appraisal.

What specific skills have you learnt which make you a good teacher?

I have learned a lot through experience, observing good teachers and the course. A few of the specific skills which I have learned and regularly employ are:

- Plan my teachings in terms of learning outcomes. This is rather obvious as the purpose of teaching is to foster learning. The purpose of my teaching is to make students think and learn rather than being a source of information and factual knowledge.

- Introduce my teaching by exploring or referring to the background knowledge (schema activation) needed for the session. Subsequent teaching would be organised and structured (schema building) so that key concepts were presented in a hierarchic order. I would summarise at the end to reinforce key concepts (schema refining). This is based on the cognitive theory principles that learning is a constructive process of schema activation, schema construction and schema refinement. This suggests that existing knowledge acts as a scaffold on which new knowledge structures are built. The learning is likely to be much more effective if prior knowledge is activated before presenting the new concepts. Also the new knowledge would be assimilated better if it were presented in a structured and organised way. The new schema could be further refined and reinforced by summarising the key concepts at the end.

- Actively encourage students to take charge of their learning and foster deep learning principles. I would provide key concepts and encourage students to search for other relevant material on the issue to develop an understanding. Another powerful underlying principle is one of behaviourist theory that independent learners learn best. Students learn best when they take responsibility for their own learning.

- Create an environment of trust, relationship and mutual respect by being non-threatening, committed and responsive to the needs and aspirations of the learners.

- Provide regular feedbacks and encourage them if they are doing well and guide them in the deficient areas. Positive feedback is important not only from the reinforcement principle of stimulus-response theory but also the learning is likely to be more effective and efficient if the learners are informed as to how well they are doing (cognitive feedback principle).

- Provide teaching in a real world context i.e. create a context in which the problem is relevant. Encourage students to construct multiple perspectives on an issue either by suitable examples or by collaborative learning. Very often learners fail to transfer classroom teaching to the real world. The teaching should thus be in a real world context and further multiple perspectives should be debated so that the learner can adopt the perspective that is most suitable to them in the particular context. Also, for the teaching to be effective, the learning of content should be embedded in the use of that content. This would avoid the difficulties of putting the theory in practice (constructivist cognition).

What one technique has had the biggest impact on your teaching methods?

Constructivist cognitive theory of education, schema activation, building and refinement.

So I introduce my teaching by exploring or referring to the background knowledge needed for the session. Subsequent teaching would be organised and structured so that key concepts are presented in a hierarchic order. I would summarise at the end to reinforce key concepts.

What methods of teaching do you know? Which do you prefer and why?

Lectures, seminars/tutorials, problem based learning (PBL), one to one teaching (e.g. learning endoscopy), small group teaching (clinical teaching or teaching ward rounds).

I prefer problem-based learning. The basic principle is that the students are not passive learners but actively learn for themselves using the problem as a focus of their learning. The students move from the problem towards the rule, principle or concept and then generalise their learning to other contexts or settings. So it is not simply the opportunity to solve problems, but rather learning opportunities where solving problems is the focus or starting point for students learning. So the advantages of PBL are:

- It is student-centered. It promotes active learning and thus improves understanding and retention and development of lifelong learning skills.

- PBL approach contributes to the acquisition of generic skills and attitudes essential for future practice.

- PBL is fun and rated enjoyable by both students and staff. It motivates students by freeing them from rote learning and use of clinical setting for the scenarios.

- PBL encourages a deep approach to learning. The students interact with the learning material, relate concepts to everyday experience and evidence is related to conclusions.

- PBL facilitates a constructivist approach to learning. When generating learning issues, students activate prior knowledge and build on existing conceptual knowledge frameworks.

You could also say that you prefer small group teaching, e.g. ward-based clinical teaching in groups.

Small group teaching is an important educational strategy. Small group learning is not defined by the number of learners although admittedly meaningful interaction occurs more readily with fewer people. Small group learning is defined by its three key characteristics: active participation, a specific task and reflective learning.

Small group learning has many advantages:

- It fosters active learning. Group discussion activates prior knowledge helping identify any deficits and facilitating new understanding.

- Small group work allows students to self-direct their own learning. It thus promotes lifelong learning.

- Small group work is more interactive and hence increases learners' involvement and thus motivates them to learn and learn more effectively.

- Small group discussion allows application and development of ideas by allowing students to explore different possibilities.

- Small group work allows for a deep-learning approach because students understand and make personal sense of the material rather than just memorising and reproducing (superficial learning).

- It promotes an adult style of learning by encouraging students to take charge of their own learning.

- Small group work fosters team working spirit, problem solving abilities and communication skills. Development of these transferable skills is important in the management of all patients.

There are various methods that can be used with small groups like tutorial, seminars, problem based learning, clinical teaching etc. I suggest you use the clinical teaching method. You could conduct it on the hospital wards around patients' bedsides. It provides real life experience of real patients and hence medical students enjoy it the most. It provides good opportunity for the observation (and correction if needed) of clinical, communication and interpersonal skills. Clinical teaching allows the integration of cognitive, psychomotor and effective objectives.

You are given a group of six core trainees to teach in a week's time on a subject to be chosen by you. How do you go about preparing for it?

You decide on the method of teaching, e.g. clinical teaching or ward-based teaching. You can demonstrate your passion and commitment for teaching by answering this question appropriately.

- I would first of all familiarise myself of the students' needs, their stage of the course and the requirements of this teaching session.

- I would chose patients who are sufficiently well to be seen by the students. I would get consent from them and brief them adequately so that they knew what will be expected of them during the session. They would be given opportunity to refuse. Further, I understand that clinical care of the patient is paramount. I would communicate with the nurses so that they were aware of the teaching plan and thus should minimise interference in clinical care or teaching.

- I would brief the students on the learning outcomes of the session.

- I would articulate a few selected teaching points per case and communicate these points through questions and discussions. I would try and link the facts of the case to the general principles of medicine. I would use an interactive style of teaching. I would involve all the students by encouraging the more reserved to participate and limiting the contribution of more vocal members. I would use appropriate questions to draw upon the prior knowledge of the students. When prior knowledge is lacking, I would offer a conceptual scaffolding and context for learning. I would stimulate interest by being challenging and buttressing the relevance of the teaching to a variety of clinical situations

- I would thank the patients and the health care team for their contribution in the clinical session. I would use the ward side room for debriefing after the session. I would also use it to discuss the sensitive issues raised earlier in the discussion. In my debriefing, I would clarify any misconceptions or misunderstandings to reinforce student learning. I would encourage students to reflect on their recent clinical interaction in the light of previous experiences. I would recommend use of a log book or a portfolio to help with this process. I would also provide constructive feedback and suggest useful further readings. Finally, I would reflect on the session myself as to how well I was able to link the experience to the students to other clinical experiences. I would also seek feedback from the students.

What would you teach a group of junior core trainees in 30 minutes?

Professionalism, good attitudes and reflective thinking.

You could also do a case-based discussion, demonstrate physical examination. Basically, you could choose anything as long as you explained the importance of it.

How would you convince a junior colleague of the importance of teaching?

Ask him about an inspiring teacher and then ask him how it influenced him.

What is Problem Based Learning (PBL)? What are its pros and cons?

The term PBL is employed to convey different concepts. The principle idea behind PBL is that the starting point for learning is a problem that the learner wishes to solve. The basic outline of the PBL process is: encountering the problem first, problem solving with clinical skills and identifying learning needs in an interactive process, self-study, applying newly gained knowledge to the problem, and summarising what has been learnt.

Pros of PBL (discussed earlier)

Cons

- PBL makes it very difficult for students to identify with a good teacher. In PBL, the teacher serves as a facilitator rather than acting as a role model. This may deprive students of the benefits of learning from an inspirational teacher.

- PBL does not motivate staff to share knowledge with the students. Staff are denied the fun of sharing their processes of understanding with their students and of 'getting a buzz' out of teaching.

- PBL require competencies many teachers do not possess.

- PBL may be time-consuming for students, particularly if they need to identify educational resources for themselves. The use of study guides will minimise this potential drawback.

- Concerns have also been raised about the cost of implementing a PBL programme.

How do you know what you don't know?

How do you identify your training needs?

What measures do you take to improve your training?

By reflecting on my practice.

The medical professional is now expected to reflect upon their practice, identify their learning needs, plan and undertake the learning and then evaluate the process.

So I

Review- look at my own life experiences.

Reflect- sort out what I have learned from these experiences.

Record- document the insights gained from taking an honest look at myself.

These steps would lead to the identification of learning needs. You will next need to think about how or what needs to be undertaken to successfully achieve the identified learning needs. This will help you identify the learning resources needed and a reasonable time scale.

I am also a self-directed learner and keep myself up to date by regularly reading journals and attending conferences/meetings.

Tell me about a memorable case where you have learnt something new?

Quote a case relevant to your specialty.

How do you assess surgical competence in a trainee?

Competence is knowledge, skills and attitudes and each of this need to be assessed to assess competence. A combination of methods like Mini CEX, DOPS, MSF, and OSCE are used to assess surgical competence. Direct observation by the trainer and assessment of log book provide further evidence. Feedback from colleagues and allied health professional is also crucial.

Do registrars have a role in teaching core trainees?

Absolutely, because registrars are directly supervising core trainees and are thus in a position to identify and fulfill the learning needs of core trainees.

With the introduction of the European Working Time Directive and the reduction in the number of hours leading to CCT, do you feel you will be fully trained by the end of your registrar post?

Yes

- Focused and competency-based training.

- Learning does not stop with the end of training. What you are saying is that although you are competent and excited at taking the next step up to become a consultant, you do appreciate and believe in lifelong learning.

- Reflective learner. This has helped you identify the gaps and lacunae in your training/knowledge/skills, leading you to take steps to develop yourself further.

How do you think trainees should contribute to their own training?

Trainees have responsibility to take charge of their own training. They need to reflect on their practice, identify the learning needs and use appropriate resources to fulfill them. They need to liaise with their supervisors and plan their learning outcomes for each stage of their training.

What do you get out of teaching others?

I get a 'buzz' out of teaching by sharing my processes of understanding with the students. It gives me no end of satisfaction when a trainee comes up and tells me that my teaching session helped him manage his pts. better. Teaching also helps me consolidate my own knowledge.

What is the most interesting case you have managed?

What is the worst case you have managed?

When did you last call your consultant?

What is the biggest mistake that you have made in a clinical setting?

Tell me about a clinical situation where you've needed to seek advice; what lessons did you learn from it?

These questions are all the same.

Discuss a case relevant to your specialty. Try and focus on a non-clinical issue like communication skill, ethics, empathy, management skills etc. This is what you would need as a consultant.

How do you keep your skills up to date?

I constantly update myself by attending meetings, reading journals etc.

How did you keep your skills up to date during you research/career break?

By attending local educational meetings as well LETB organised ones besides reading journals. I also attended one clinic and one theatre session a week.

Tell me about the most recent paper you've read which will change your day to day clinical practice.

Tell me about an interesting paper you've read in the past three months.

Topical question relevant to your specialty.

What invasive procedures have you performed and what complications have you encountered?

Most of us will have encountered complications. What is important in answering the question is how you dealt with it. Important steps

- Took steps to ensure patient safety
- Involved your consultant immediately
- Explained everything to the patient as soon as possible
- Reflected on it and identified learning points

If you could improve the specialty training scheme in one way, what would you do?

Be honest and narrate your reasons for it.

You could say that you would like learning outcomes to be set for each stage of training.

What do you understand by 'personal portfolio'?

A portfolio is a purposeful collection of work that when put together, demonstrates that learning has taken place. The portfolio's purpose is to demonstrate learning and not to chronicle a series of experiences.

What is the role of the LETB (previously known as deanery)?

The LETB are responsible for the management and delivery of postgraduate medical education and for the continuing professional development of all doctors and dentists. This includes ensuring that all training posts provide the necessary opportunities for doctors and dentists in training to realise their full potential and provide high quality patient care. The LETB are also responsible for trainers, educational supervisors and educational leaders, their training needs and educational development.

PS: Under the Health and Social Care Act 2012, the role of the deanery has been taken over by Local Education and Training Board (LETB)

Does formal qualification on medical law or education or management make you more effective as NHS consultant? If so, how?

There are two issues here: formal qualifications and experience. Clearly, having a formal qualification does not guarantee competence. However, going through a formal qualification requires a degree of study in the subject and usually results in learning a lot. Experience alone can be an uncertain indicator of quality.

Obviously, in the end, it's the capability that matters. However, experience, track record and formal qualifications are the only real evidence that can be looked at to judge whether someone is effective and capable.

(NB- You can emphasise the importance of experience and track record if you do not have formal qualification. After all, most presidents and prime ministers do not have formal management qualifications and managing the country is probably one of the biggest jobs out there. Similarly, if you do have formal qualifications, you can quite rightly mention you would not fly if the pilot of the plane did not have formal qualifications.)

Ethical Issues and Difficult Work Scenarios

How would you handle a problem doctor - for example if you suspected that your consultant had a drink problem? What would you do if you found a colleague taking illicit drugs?

How would you react if one of your junior colleagues turned up drunk on the ward first thing in the morning? What if it were your consultant? (NCAS advice below)

A patient mentions to you that on two occasions they have smelt alcohol on your consultant's breath during clinic in the past few weeks. What do you do?

A patient mentions that, during an examination, one of your colleagues examined her breasts. Although the patient is not aware that such behavior was inappropriate in that context, you are. How do you respond?

You observe your consultant making inappropriate sexual remarks to one of your patients. There are no other witnesses and the consultant is not aware that you were there. How do you react?

You have heard rumours that one of your colleagues is taking drugs. You also know that some drugs have disappeared from the cabinet. How do you react?

You have suspicions that one of your peers has been stealing an important amount of hospital property (including stationary and needles). What do you do?

These questions are the same, dealing with significant concerns.

Dealing with a drunk colleague:

My first concern would be the safety of the patient. However I have a duty of care to my colleague and the hospital too.

Patient safety: I will talk to my colleague/consultant and send them home. (If difficult, enlist the help of another consultant or clinical director). I will review all the patients seen by him and complete the ward round. I will also recall all the patients discharged by them. I will ensure that appropriate cover is arranged for them, if needed.

Duty of care to colleague: I will arrange a taxi for them to go home and check on them later to ensure they have reached home safely. I will discuss with them (if a consultant I will pass on the incident to a clinical director) and try and help. You could suggest occupational health referral. Insist that his behavior was not appropriate.

Duty of care to the hospital. Keep accurate records and inform your consultant or clinical director.

Background

You must protect patients from risk of harm posed by another colleague's conduct, performance or health. The safety of patients must come first at all times. If you have concerns that a colleague may not be fit to practice, you must take appropriate steps without delay, so that the concerns are investigated and patients protected where necessary. You have a few options:

- You can report your concerns to your clinical or medical director.
- You can discuss your concerns with National Clinical Assessment Service (NCAS).
- Report your concerns to the GMC.

If you think the action taken has been insufficient, contact the GMC. Acting on concerns about a colleague is never easy but all NHS staff have a professional duty to do so in order to protect patient safety and help the practitioner involved.

Where concerns about performance have arisen it may be helpful, at any stage of the process, to consider why this has happened. THINK ABOUT:

The individual's health and other factors

- Does the individual have a physical or mental illness?
- Is the individual depressed or suffering from another mental illness?
- Might alcohol or substance misuse be involved?

- Has there been a recent major life event?

Knowledge, skills and behavior

- Is there a difficulty with clinical knowledge and skills?
- Might a deficiency in education, supervision or continuing professional education be contributing to the problem?
- Was the practitioner's induction appropriate or sufficient?
- Does the individual have difficulty understanding the limits of their competence?
- Is the problem predominantly one of the practitioner's behaviour or attitude?
- Is this new behaviour or is it an exacerbation of long-standing problems?

The job

- Have work factors changed?
- Is there a problem with technological advances or techniques?
- The work environment
- Are there team difficulties?
- Have there been major organisational changes?
- Could issues relating to equality and diversity be a problem?
- Could bullying or harassment be a problem?
- Are there any systems issues that contributed to the performance difficulty?

What is meant by a significant concern?

Significant concerns about a practitioner may relate to any of the following areas:

- Poor clinical performance

- Ill-treating patients
- Unacceptable behaviour such as harassing or unlawfully discriminating against staff or patients
- Breaching sexual or other boundaries with patients or staff
- Poor teamwork that compromises patient care
- Personal health problems leading to poor practice
- Not complying with professional codes of conduct
- Poor management or administration that compromises patient care
- Suspected fraud or criminal offence

This list is not exhaustive and there may be other areas of concern that you should consider reporting.

How would you react if one of your female junior colleagues refused to treat a patient who is a known rapist?

GMC guidance: If carrying out a particular procedure or giving advice about it conflicts with your religious or moral beliefs, and this conflict might affect the treatment or advice you provide, you must explain this to the patient and tell them they have the right to see another doctor. You must be satisfied that the patient has sufficient information to enable them to exercise that right. If it is not practical for a patient to arrange to see another doctor, you must ensure that arrangements are made for another suitably qualified colleague to take over your role.

All patients are entitled to care and treatment to meet their clinical needs. You must not refuse to treat a patient because their medical condition may put you at risk. If a patient poses a risk to your health or safety, you should take all available steps to minimise the risk before providing treatment or making suitable alternative arrangements for treatment.

How would you react if a patient refused to be treated by one of your junior doctors because he was foreign? You see a patient verbally abuse a nurse. What is your response?

Firstly, ensure that the patient's behavior is not due to his/her underlying medical condition. If it is willful, warn the patient that the behavior is unacceptable. If the patient persists, he should be given a formal warning. If the behavior is repetitive, he may be removed from the trust premises by the hospital security. However, his care should not be hampered and the responsible clinician should make adequate arrangements for transfer of his care.

Most trusts have policies regarding dealing with violent/abusive/racist patients. Such behavior is considered an assault on a member of staff under many trust policies. Most trusts have a zero tolerance policy towards abusive patients and care may be withdrawn from persistent offenders.

Most trust policies would recommend

- Explain to the patient that their behaviour is unacceptable and explain the expected standards of behaviour, which must be observed in the future.

- If the behaviour continues, the responsible clinician will give an informal warning about the possible consequences of any further repetition.

- Continued behavior after a formal warning will lead to immediate exclusion from the trust premises by the security staff/police. Such an exclusion from trust premises would not mean that he would not receive care, as his clinician would make alternative arrangements for him to receive treatment.

Tell us about an ethical dilemma you have been involved in.

Remember, medical ethics rest upon four key principles:

- The principle of autonomy – individuals have a right to be self-governing

- The principle of non-maleficence – the patient should not be harmed

- The principle of beneficence – the benefit of the patient should be promoted

- The principle of justice – equals should be considered equally

Two representative examples of ethical dilemmas:

Mary is clinically depressed and takes a lethal overdose. She leaves written instructions asking not to be resuscitated. If you arrived at Mary's side in time to do so, would you resuscitate her?

Living wills (advance directives) are valid in English Law. However, an advance directive must be written by a person who has the mental capacity. We do not know if Mary had psychotic depression. Nor do we know if she understood the consequences of her refusal of future treatment. The clinician should treat Mary until she is sufficiently competent to make her own decision about further treatment. The legal defense of 'necessity' would cover any treatment (including resuscitation) which was necessary to protect Mary's life, and which could not reasonably be delayed.

The idea of this question is to discuss an ethical dilemma. The most important bit of the answer is how you dealt with it. So in the above question, you could have discussed with the psychiatrist to get further guidance. (So define the dilemma and then discuss how you resolved it)

Another example

Mike is in a persistent vegetative state. Mike is legally alive (spontaneous respirations and heartbeat) but a decision to withdraw nutrients and antibiotics with the intention of ending his life would not necessarily be unlawful.

There is no legal duty to provide treatment that is no longer considered in the patient's best interests. Although the withdrawal of life support is factually the cause of death, the legal characterisation of this 'terminal regime' as acting in the best interests of the patient, means that the causative action has no legal consequences.

The GMC has launched a series of challenging online tutorials that tackle tricky ethical scenarios.

Visit the interactive case studies at http://tinyurl.com/6ya7jxk

One of your peers arrives constantly late for work in the morning. What do you do?

For minor concerns (coming late) about performance, an informal approach may be all that is needed. Here, a discussion with the individual concerned, aimed at improving their performance or conduct, may be sufficient to resolve the issue. Dealing with the matter informally provides the opportunity for both parties to agree the way forward without the use of formal disciplinary or other procedures. Even if an informal approach is taken, the outcome of the discussion and agreement reached should be communicated to the practitioner in writing and notes kept of all meetings held.

Your consultant does something that goes against protocol. How you do you tackle it? How would you approach the consultant?

Your consultant mentions something to a patient that you believe to be wrong. How do you react?

Don't assume that your consultant is wrong. Discuss with him that you found his decision interesting and would like to learn the thinking behind it. If you are not satisfied and think that less than perfect care has been provided, you are duty-bound to raise your concerns with your consultant. If you find it difficult, discuss it with your supervisor or the clinical director.

Your consultant does not provide adequate training and adopts a condescending attitude towards you because of your apparent lack of knowledge. How do you react?

Your core trainee mentions that another core trainee is complaining about the fact that their consultant does not provide adequate teaching. How do you respond?

There may be reasons for the consultant's behavior, e.g. health or work related reasons. Discuss it with him and see whether you could arrive at an acceptable training solution. The problem may be with the trainee too. If there are still concerns, discuss it with the college tutor or the clinical director.

One of your colleagues seems to be suffering from stress. What do you do?

Discuss it with your colleague if you can and suggest self-referral to occupational health. If they are not willing to cooperate and if you suspect it is compromising patient care then discuss it further with your clinical director.

What is your opinion about accepting gifts?

All codes of practice for healthcare professionals have a similar rule. Note that accepting gifts is not prohibited, provided that the gift is not seen as an inducement.

In practice, this can be difficult to demonstrate, so it is probably unwise to accept any gift of significant value. 'Significant' is also a matter of judgement, of course, but it is unlikely that the code was intended to apply to the nurse who receives a box of chocolates from a grateful patient.

What do you think about choice in the NHS?

The NHS was set up principally as a public service, aiming to meet common needs rather than consumerist demand.

There is no way logically for choice to be the dominating principle of healthcare provision unless there is an infinity of healthcare provision and one single consumer...you can choose a doctor you like but if the clinics are full, choice is limited

What do you think about presumed consent for organ donation?

Although 90% of the UK population is in favour of organ donation, only 24% has signed the Organ Donation Register. Currently, when a person's wishes are not known, relatives are asked to decide about donation, in the most difficult circumstances, when they are recently bereaved. Not surprisingly, a large number of families—around 40%—opt for the default position, which is not to donate.

BMA have supported a system of presumed consent. The system of presumed consent may increase the organ availability but would still retain a role for relatives, opting out would be easy and accessible.

- One of the major concerns people have with a presumed consent system is that individuals will lose control over what will happen to their body after death, and the state will take over. This is not the case. Like the current system, under presumed consent people would retain the choice over whether or not to donate after death.

- Mechanisms must be in place to ensure all members of the public are informed of their choices and can register an objection quickly and easily—for example, through their general practitioner, post office, or electoral registration forms. As an added safeguard, the system would retain a role for relatives. After death, relatives would be informed that the deceased person had not opted out of donation and, unless they object—either because they know of an unregistered objection by the person or because it would cause major distress to the close relatives—the donation would proceed.

- The opt-out proposal will not mean that those who do not wish to donate their organs will have to do so, or that families will not have a choice. What it will mean is that everyone will be prompted to think about that choice, to make a decision and discuss it with their loved ones, rather than avoiding the issue and thinking, as is all too easy to do...

Author's opinion:

However, I believe presumed consent is not consent at all. The Human tissue act rightly puts consent at the heart of the act for the removal and use of human organs. To increase the organ supply, we should mandate all adults to make a choice regarding organ donation. A mandated choice will help quickly resolve the issue.

How would you handle a non-performing junior/consultant colleague?

A poorly performing doctor is someone whose competence, conduct or behaviour poses a potential risk to patient safety or to the effective running of a clinical team.

A "problem" junior is someone who does not meet expectations due to deficiency in one of three areas: knowledge, attitudes, or skills.

- Knowledge deficiencies in basic or clinical sciences.

- Attitude problems (usually manifested as behaviours) - difficulties related to motivation, insight, doctor-patient relationships and self-assessment.

- Skill deficiencies include problems with interpersonal or technical skills, or clinical judgment and organisation of work.

Steps in resolving/handling a non performing colleague

Diagnosis

- Assess the full details of the problem- how long and how bad. Identify the problem with the junior- is it knowledge, skills or attitudes?

- Discuss with the learner to find out whether he has insight into the problems and whether there are any external factors like life stresses (divorce, immigration etc), substance misuse, psychiatric illness etc which may be causing the problem.

- Assess whether there are any system problems like excess workload, inadequate teaching and training or lack of feedback.

Confirming the diagnosis

This is done by careful collection of data by

- Direct observation of the problem doctor in a variety of situations
- Feedback from other colleagues
- Feedback from other rotations in other specialty and hospitals

Intervention

Once the problem is diagnosed, we need to determine how we will intervene, who should be involved, and when to evaluate outcome. The intervention will vary depending on the problem diagnosed. Common interventions include:

- Communicate clear expectations
- Provide enhanced teaching and learning opportunities
- Arrange for peer or mentor support
- Reduce the clinical workload, with more protected time
- Recommend counselling and/or therapy

Bottom line is that you must deal with the problem positively by involving more senior people. If they cover it up then involve even more senior people. Remember, if you know about something and do nothing, you are in just as much trouble if an inquiry takes place.

One of your junior colleagues is placing patients at risk. How do you react?

The patient's safety is paramount. So notify immediately his supervising consultant. The other steps are as in the answer above to dealing with a non performing colleague.

Management Questions from the chief executive

You will be required to balance both outpatients and inpatients demands, taking into account government targets. How will you go about this?

Describe how you would manage the diverse demands on the service given the resources available to you.

How would you manage the budget?

All these questions are essentially about setting a budget and managing it.

A budget is essentially a financial plan for the short term, usually one year, allocated to each department. The budget is divided between pay or fixed (staff related expenditure) and non-pay or variable expenditure (goods and services) to cover all the running costs of the department for the duration of the budget.

Setting a budget: This depends on forecasting the number of patients, patient days, nursing resources needed, other costs of supplies and personnel etc. Decisions on setting the budget should be based on reality and real trends rather than historic data.

The process of managing the budget can be broken down into 4 stages:

- Establish the actual position: In general, the budget can be divided into monthly blocks. It is important to focus on the future position to control the budget. The calculation of out-turn (amount of spending to date/number of months to date X 12 equals projected out-turn) formula will provide the year end position for pay and non-pay expenditure.

- Compare actual expenditure with budget totals: this will indicate whether the budget is over- or under-spent and help you identify the spending pattern. For example, use of bank and agency staff will create potential overspending, thus careful monitoring is essential.

- Establish reasons for variance: Variance is the difference between the budgeted amount for the month and the actual amount spent. The reasons for variance must be sought. It could be due to an anticipated increase or decrease in workload and therefore of no particular concern. However, variance may be completely unexpected and in such cases the reasons must be found.

- Take action: Variance could be due to mistakes. Items may have been wrongly attributed to a budget or miscoded and end up in the wrong division of the budget. These problems can be corrected with the help of the finance manager. The budget must be checked for every transaction and corrected where necessary.

How will you spend 50k on service improvement in the department?

Your pre-interview visit will give you a good insight as to the need for new services or any improvement needed in the services provided. Discuss the need for the service, the benefits it will bring (try and marry it to NICE guidance, waiting targets or the trust's strategic direction, if possible) and the costs involved (i.e. 50k). Don't forget to mention you will talk to all stakeholders so that they have ownership of the service. Also need to mention clinical governance issues, i.e. regular audit to ensure the service is meeting expectations.

Tell me about changes in practice that you have helped initiate in your posts so far?

Discuss any changes you made in your training post. Most changes usually happen as a result of audits. Discuss your experience of change management e.g. talking to colleagues, bringing a difficult person on board, arriving at a consensus etc. Discuss also how the change was implemented and how it led to improved services.

You may also discuss any hospital guidelines authored by you or changes to the rota initiated by yourself.

Change management

RAID is a 4 step process to manage change:

R- Review- where are we now? Gathering information, listening to pts. and staff, looking at audit, documentation and process.

A- Agree- gain consensus, build teams, formulate recommendations, shape the future.

I- Intervene- project management, priorities, dealing with transition, motivate and support staff, expect and deal with resistance.

D- Demonstrate- project analysis, show the differences, identify lessons, plan next objectives.

Tell me how you would bring this new technique into the trust.

You may be required to set up services not presently in existence. How would you go about this?

How will you make sure patients play a part in the set-up of the service?

The steps to setting up a new service/technique are: defining the need, costs and benefits of the service, funding, writing a business case and implementation.

Defining the need

Why are new services needed? Will the service alleviate a significant risk to the trust? Will the service help the trust and/or commissioners meet government targets? Does the service fit in with the trusts strategic direction?

Demand and capacity theory is at the forefront of a lot of modernisation work. You will need to demonstrate that the demand on your service exceeds its capacity, whether that is staff, buildings or equipment. Activity data is a powerful tool and most trusts will have an Information Team who would be able to provide you with a range of activity information linked to you or your service.

Costs and benefits of the services

Cost-assess the resources required including staff. You will need to discuss the funding with potential sponsors. Sponsors could be your own trust or the CCG.

Funding

What type of funding might I access?

There are two types of funding: revenue (this funding is added to your budget year on year; for example, staff salaries have a revenue implication) and capital (this funding is a one-off resource allocation and is usually linked to equipment or buildings).

Where would the funding come from?

Trust Capital or from commissioners. Various incentive schemes - can provide useful funding opportunities. This type of scheme will usually have a proforma which will tell you the information that will be considered in the bidding process - it will also be explicit in what it requires to be delivered in return for the investment, particularly around waiting times and access.

Business case

Write up a business case, stating basically the need for the services (detail your research) and the cost and benefits of the services. Be clear about both costs and benefits for your department, the organisation, the wider community and include immediate, short and long term implications. It is often the financial implications of proposals that carry the most weight and so make the financial case for the changes you are proposing as powerfully as possible.

Your service or general manager will help you write up the business case. They are there to support you in delivering your clinical service. The service manager and/or a member of the financial team can help convert your ideas into the required business case.

Implementation

Implement the changes gradually with strict monitoring. Audit the service to ensure that the objectives set in the business plan are achieved.

Public and patient involvement is vital in the development and improvement of any services. Services need to be responsive to the needs of the patients and public as they are end users of all services. I would ensure patients' involvement in services by involving the public and patient forum as well as regular patient surveys.

Case Study 1 - The Clinical Nurse Specialist

Problem: busy clinics, long waiting list and perhaps an NSF requirement.

Solution: introduction of a clinical nurse specialist to work alongside consultant teams in clinic and undertake a telephone follow-up clinic.

In your business case you will need to:

- establish the case for need
- provide an analysis of current and predicted demand
- assess how the proposal will resolve future service demands (activity flow changes, operational policy changes)
- estimate costs of the proposal - capital or revenue - e.g. salary, equipment, accommodation etc.
- provide the associated support costs e.g. perhaps an increase in diagnostics.

Case Study 2 - Medical Equipment

Problem: equipment at the end of its asset life and/or opportunities for advances in equipment to support service provision.

Solution: replacement of equipment with a revised specification.

In your business case you will need to:

- Establish the case for need

- Provide options for purchase including changes in practice and operational requirements
- Cost assessment - total purchase or lease, consumable costs, staff changes and maintenance requirements
- Clinical risk assessment and quality advantages
- The implications of 'do nothing'.

What do you mean by Service Line Reporting (SLR)?

- SLR means departments are accountable for their income and expenditure. This means that departments will be charged for all the resources it uses i.e. every blood test, every imaging or every cross specialty consult.
- The idea is to use resources efficiently to ensure expenditure of departments do not exceed its income.
- Watch out: Some departments may underuse resources (to balance books) to the detriment of patients. Proper governance structures and outcome data will be needed to avoid any patient consequences.

A major incident occurs and you are the consultant on call. How do you handle it?

You are the consultant in casualty. A police officer informs you on a weekend that there are 100 casualties in a football ground. How will you deal with it?

These two question deal with the same theme.

I would promptly assess the situation and discuss the situation with the clinical/medical director or chief executive and initiate the major incident policy of the trust. Every trust has a major incident policy.

A major incident is any occurrence that presents serious threat to the health of the community, disruption to the service or causes (or is likely to cause) such numbers or types of casualties as to require special arrangements to be implemented by hospitals, ambulance trusts or primary care organisations.

What is a major incident plan?

- To describe how the Trust operates in the event of a major incident (internal or external)
- To assist staff, by providing a framework for action, specific instructions and resources
- To ensure that incident responses are structured, co-ordinated and managed effectively from the outset
- To enable the Trust's response to be co-ordinated with others
- To ensure appropriate communication channels
- To enable effectiveness of actions to be evaluated

Staff morale can sometimes be low. What do you intend to do to help this?

- Treating them as human beings
- Valuing their contribution
- Feedback
- Involving them in decision making
- Understanding their aspirations
- Providing effective leadership

Give an example of where you've prioritised clinical need?

As registrars on call, we all prioritise clinical needs. Mention any particularly busy on call and tell them how you marshaled your team to effectively manage the team.

- Requested the charge nurse to triage the patients
- Minor cases allocated to junior members

- Serious cases seen by core trainee with my support
- Kept my consultant informed
- Regular feedback to the team members

What is your take on litigation?

Well, I am a good doctor. But I expect to be involved in litigation because it is likely to happen as it comes with the territory because of the complexity of the cases and interpersonal relationships.

You are the consultant on call. You have been informed that an inpatient with varices is re-bleeding and there have been two admissions in the last few minutes with complete heart block and respiratory failure needing BiPAP. The registrar is panicking. What do you do?

Clinical scenarios are common at interviews, however at a consultant interview, the focus is not on clinical medicine, but towards people management and governance.

The answer needs to show how the candidate would take charge there and then, but also that they know the problem does not end there. He/she has to take reasonable steps to prevent the recurrence of the situation.

It would be better for the consultant to say that he would form a plan with his registrar and phone his consultant colleagues. So for example, a plan may be that he seeks anaesthetic help to deal with the respiratory failure. Ask the core trainee to start isoprenaline drip and the registrar deal with the bleed with SB tube as needed or call in GI colleagues if available. You will need to go in and help out.

The consultant should debrief the juniors, discuss what happened with the clinical director, and discuss it in the clinical governance meeting.

What is your opinion of a sub-consultant grade?

Successful completion of higher specialist training as confirmed by assessments of knowledge, skills and attitudes will lead to a certificate of completion of training that confirms readiness for independent practice in that specialty at consultant level.

It is vital that training and the level of a certificate of completion of training is maintained for public and patient confidence, efficient delivery of service and for the international reputation of UK medicine.

The final years of higher specialty training should be used to prepare for consultant practice and not to bring in a new grade. The proposed new grade looks to the past with rose tinted glasses at the old senior registrar grade rather than forward to the future and how high quality healthcare can be provided by fully trained doctors. As expansion slows with respect to our consultant workforce, the current pyramidal structure may no longer be sustainable. Creating a new grade will only create a new bottleneck, not a solution to poor workforce planning.

Discuss the new GMC guidance on leadership and management.

The GMC published new guidance in March 2012 on leadership and management for doctors that spells out their responsibility for the safety and wellbeing of patients when performing non-clinical duties.

The guidance sets out the wider management and leadership responsibilities of doctors in the workplace, including in relation to employment issues, teaching and training, planning, using and managing resources, raising and acting on concerns and participating in service improvement and development. The principles in the guidance apply to all doctors regardless of whether they work directly with patients or whether they have a formal management role.

Doctors are still accountable to the GMC for their non-clinical duties, such as their behaviour as a manager and the guidance states.

The guidance states that being a good doctor involves more than simply being a good clinician. It involves commitment to improving the quality of services and to demonstrate leadership in managing and using resources effectively.

The guidance also contains a section on allocating resources, in particular exploring some of the areas that doctors will need to think about when resources are limited, including:

- Considering the needs of patients AND the wider population.
- Being familiar with local and national policies on access to treatments.
- Making sure that decisions about access to care are based on clinical need and the likely effectiveness of treatments.
- Being honest with patients about the decision making process.

Ref: www.gmc-uk.org/leadership.

The government's plan outlined in "Care closer to home" is to move more patient care into the community. Do you think this is a good thing?

Care closer to home provides more convenient and accessible services. It is thus a laudable concept, provided that high levels of quality and service are maintained.

Do you think the 2 week rule referral system is effective?

The idea behind the two week referral is good. However, a lot of 2 week referrals do not follow the DoH guidance. These guidelines may increase diagnostic precision if adhered to rigidly. Inappropriate referrals have extended already lengthy outpatient waiting times in many specialties. Furthermore, the studies have shown that the cancer pick-up rates is no higher in the 2 week pathway. There is also criticism that fast track referral prioritises the worried well at the expense of the target population. So in summary, we need better predictors of cancer and better pathways. Reduction in NHS waiting times and implementation of 18 week pathway are steps in the right direction.

How can we persuade the public that doctors can be trusted?

- Strong clinical governance framework.

- Structured and streamlined training of doctors.

- Appraisals and revalidation to ensure the public could be confident that poorly performing doctors were being identified and early action taken in order to protect patients.

- Promoting an open culture in healthcare to ensure mistakes and near misses are reported and discussed.

- Patient and public involvement in designing clinical services.

- Impeccable integrity and good role models.

- Increased involvement of public in the regulatory bodies like GMC.

Is the expanding role of nurses a benefit or a danger to the medical profession?

What is your view of nurse specialists?

A greater use of skill mix is needed with the implementation of EWTD. So tasks traditionally undertaken by doctors will need to be done by other allied health professionals. I fully support this provided that:

- Tasks are only be undertaken by individuals who are competent to perform them

- Such individuals are permitted to make decisions within the scope of their professional practice but otherwise need to operate under clear protocols and accountability.

How would you develop the current clinical service? What do you hope to achieve in the first year, if you are successful in being appointed?

How you can contribute to service (more than another candidate)?

Your pre-interview visit will come in handy here. Identify their service needs and relate it to your special skill or experience.

What could you do to improve the organisation and running of your current workplace environment?

Again, at your pre-interview visit, you may have identified a few things that need improvement. Discuss them without criticising.

What is your opinion of performance related pay?

Performance related pay means extra financial inducement for personnel who achieve certain targets e.g. throughput of patients/reduction in waiting lists/improvement in quality of patient care etc.

Advantages

- Inducements that can allow targets to be met
- Popular with workers

Disadvantages

- Not evenly spread. E.g. consultants/managers get inducements for ops, but not nurses, ODA's and theatre support staff.
- Potential resentment between groups and between hospitals.

What do you think about the NHS's association with industry – links between NHS and industry?

Advantages

- Increase quality of care for patients

- Mirror the use of some techniques in industry to motivate staff

- Increased new research

Disadvantages

- Size of company – their influence may be "pervasive"

- Potential conflicts of interest when clinicians are funded for study leave research grants

- Other financial payments

- Potential to influence CCGs

What do you think about management issues? Do you think it's something you should be getting involved in as a clinician?

Clinicians serve the public and the patients by using their skills to provide the best possible advice, treatment and care. But we can only do this if the money available to the NHS is used well. Failure to do so results in less care and of a lower quality. Money will only be used well if clinicians are fully engaged in managing it. Ultimately, it is clinicians who are responsible for the way in which services are delivered to individual patients and it is they who commit the necessary resources.

Improving the quality of care and providing more responsive services for patients can only be achieved by strong involvement of local clinicians in the management of the service.

This includes having the understanding, the tools and the ability to manage resources effectively and use them well to the benefit of patients. This will empower them to lead change and improve services. Without clinical involvement, the progress will be much slower and the outcomes poorer.

This is not about focusing on cost and cost alone. It is about how money can best be used to improve the quality of care, combining operational and clinical effectiveness. Efficient use of resources and good quality services go hand-in-hand.

Clinical Governance

What do you know about clinical governance (CG)?

Clinical governance is the framework that helps organisations provides safe and high quality care. It is fundamentally about the ability to produce effective change so that high-quality care is achieved.

This definition is intended to embody three key attributes: recognisably high standards of care, transparent responsibility and accountability for those standards, and a constant dynamic of improvement.

Clinical governance relates to only those aspects of such organisations that relate to the delivery of care to patients and their carers; it is not concerned with the other business processes of the organisation except insofar as they affect the delivery of care.

The chief executive is responsible for its implementation. He/she delegates most of the tasks to other colleagues like the director of nursing, medical director etc. Ultimately, each healthcare professional has a role to play in its implementation.

There are 7 pillars of clinical governance:

Education and training- it covers the support available to enable staff to be competent and up to date. Professional development needs to continue through lifelong learning. In NHS trusts, the continuing professional development of clinicians has been the responsibility of the trust. It has also been the professional duty of clinicians to remain up-to-date.

Clinical audit- is the review of clinical performance, the refining of clinical practice as a result and the measurement against agreed standards- a cyclical process of improving the quality of clinical care.

Clinical effectiveness- by use of evidence based medicine. Clinical effectiveness has been promoted through the development of guidelines and protocols for particular diseases.

The development of NSFs and NICE are further attempts to improve the responsiveness of the service to evidence of effectiveness.

Risk management- it involves having robust systems in place to understand, monitor and minimise the risks to patients, staff and the organisation and to learn from mistakes and past experience.

Research and development- Good professional practice has always sought to change in the light of evidence from research. The time lag for introducing such change can be very long. Techniques such as development of guidelines, protocols and implementation strategies are all tools for promoting implementation of research evidence. NSF and NICE are supporting this further.

Openness - Poor performance and poor practice can too often thrive behind closed doors. Processes which are open to public scrutiny are an essential part of quality assurance. Open proceedings and discussion about clinical governance issues should be a feature of the framework. As part of openness, the NHS publicises complaints procedures to patients, deals with problem doctors openly, encourages doctors to admit their own mistakes as part of a blame free culture etc.

Patient and public involvement- This means listening to what the public thinks of the services provided, and learning from their experiences.

Isn't clinical governance only for doctors, nurses and allied health professionals?

Definitely not! Clinical governance is for everyone. It relates to all people who are involved in the treatment and care of patients. Clinical governance is the appropriate governance (or control) of clinical (patient health) care, and therefore encompasses everything that has any impact on the quality and patient experience of that care. This means that it ranges from issues of cleanliness, comfortable environments, good food and excellent communication through to the skills required to produce first-class results in complex operations.

Similarly the domestic worker on a hospital ward has a key role to play in ensuring a clean and pleasant environment for patients, as well as helping to control infection. Other NHS staff, such as office-based managers and secretaries also contribute to high standards of care. They may never see a patient during their working day, but they have a responsibility for making certain that all aspects of their work take account of the quality and safety of care to patients.

So all NHS staff, whatever their role, are responsible for making sure that their own contribution to the massive jigsaw that makes up health care provision is delivered to the highest possible standard for patients – and that's what clinical governance is all about.

What does clinical governance mean to you?

Clinical governance to me means providing the best care possible to the patient sitting in front of me. Apply the seven pillars of clinical governance to your particular situation.

Have you seen clinical governance in action?

Yes- Quote any implementation of change you have noticed, e.g. after an audit.

How does clinical governance affect patient safety?

Discuss the seven pillars above.

Do you think the current system of clinical governance is effective?

Although the current system has made effective improvements, there is still room for further improvements for e.g. in promoting a blame free culture.

How does clinical governance impact your daily work?

Clinical governance:

- Provides a better experience for staff
- Provides a better experience for patients and carers
- Improves the quality of care for patients
- Makes the changes you want to make happen

Does clinical governance work?

Clinical governance works as long as it is:

- an active pursuit driven by, and focused on, quality
- recognises the complex nature of health care systems
- is not strangled by paper and procedure

Do you think clinical governance is useful or is it just another layer of bureaucracy?

It is essential for improvements in patient care

Clinical governance brings together all the activity that contributes to the clinical service provided to patients.

Who, in your hospital, is responsible for clinical governance?

Most trusts have a clinical governance committee headed by usually the chief executive or the medical director.

What is clinical risk management?

Clinical risk management means identifying the risks to patients and what needs to change to improve safety. It builds on the basic concept of duty of care to provide high quality healthcare protecting patients and staff. Clinical care pathways or protocols have been widely developed in some disciplines and may minimise the chances of adverse incidents occurring.

What is a near-miss situation?

A near miss is an unplanned event that did not result in injury, illness, or damage - but had the potential to do so. A critical incident is defined as one which led or could have led to harm if it was allowed to progress. It should be preventable by a change of practice.

Avoid the trap of minimising the problem: we can often learn from near-misses (where less distress occurs) better than instances where patient care is affected.

E.g. - A nurse programmes a computerised syringe pump incorrectly and it gives a patient too much of the drug too quickly. What factors do you think may have resulted in this error?

Having reviewed the incident it is vital to learn from what happened and implement preventative measures or a safety net.

In the syringe pump example, it may become obvious that it is not the most suitable apparatus: the pump design makes it complicated to use; the operating instructions are poor; and extensive training is required to achieve the required competency for use. An outcome may be to notify all potential users of the hazard and to recommend a change in the type of pump used.

What happens to critical incidents forms once they have been submitted?

It is essential to recognise that clinical incident reporting is not part of a disciplinary mechanism, but is part of a reflective system that supports an improvement in the quality of service that the directorate provides. A robust clinical incident reporting procedure in line with education and training acts as a foundation to the directorates' clinical governance framework.

By reporting and subsequently assessing any clinical incidents, clinical risks can be reduced or managed, and the overall quality of patient care improved.

A clinical incident is defined as anything associated with the patient and his or her clinical treatment or care, and is an event that led to harm, or could have led to harm if it had been allowed to progress. It should be preventable by a change of practice. It does not relate to any incidents involving staff, relatives or visitors to the Trust.

Further, these critical incident forms feed into NRLS (read the section on NRLS).

How would you introduce change?

RAID is a 4 step process to introduce any change:

R- Review- where are we now? Gathering information, listening to pts. and staff, looking at audit, documentation and process

A- Agree-gain consensus, builds teams, formulate recommendations, and shape the future

I- Intervene- project management, priorities, dealing with transition, motivate and support staff, expect and deal with resistance

D- Demonstrate- project analysis, show the differences, identify lessons, plan next objectives

What are the GMC good medical practice guidelines? What do you think the most important aspects of the GMC's 'Good medical practice' guidelines are?

Good Medical Practice sets out the principles and values on which good practice is founded; these principles together describe medical professionalism in action.

GMC good medical practice (2013) guidelines

Domain 1: Knowledge, skills and performance

- Develop and maintain your professional performance
- Apply knowledge and experience to practice
- Record your work clearly, accurately and legibly

Domain 2: Safety and quality

- Contribute to and comply with systems to protect patients

- Respond to risks to safety
- Protect patients and colleagues from any risk posed by your health

Domain 3: Communication, partnership and teamwork

- Communicate effectively
- Work collaboratively with colleagues to maintain or improve patient care
- Teaching, training, supporting and assessing
- Continuity and coordination of care
- Establish and maintain partnerships with patients

Domain 4: Maintaining trust

- Show respect for patients
- Treat patients and colleagues fairly and without discrimination
- Act with honesty and integrity

The Good medical practice (2013) was updated in 2014, to include a new duty about doctors' knowledge of the English language. 'You must have the necessary knowledge of the English language to provide a good standard of practice and care in the UK'

What is the difference between a protocol and a guideline?

A protocol is a step-by-step outline for undertaking a specific task. They normally have to be followed exactly, whereas with a guideline the recommendations need to be considered in the light of the particular patient and settings as well as the strength of the evidence base.

A guideline is 'a systematically developed statement to assist decisions about appropriate healthcare for specific circumstances.' Clinical guidelines are based on the best available evidence and provide recommendations for practice about specific clinical interventions for specific patient populations.

Who is a good doctor?

GMC- Good doctors make the care of their patients their first concern: they are competent, keep their knowledge and skills up to date, establish and maintain good relationships with patients and colleagues, are honest and trustworthy, and act with integrity.

How do you seek informed consent for the procedures that you do?

I work in close partnership with my patients. I consider consent an important part of the process of discussion and decision making, rather than an isolated process. I explain to them their condition and treatment options in a way they can understand. The details of information I share with them depend on their wishes. The information I share is in proportion to the nature of their condition, the complexity of the proposed investigation or treatment, and the seriousness of any potential side effects, complications or other risks. I respect their right to make decisions about their care. (GMC guidance)

What do you consider to be sufficient level of information for the patient?

In practice, the law has tended to regard the level of information required to be that acceptable to 'a responsible body of medical opinion' (the Bolam test). More recently, the courts have said they will depart from this approach if they see fit. The ultimate legal test being what the court itself thinks is a reasonable amount of information. The difficulty for the clinician, of course, lies in foreseeing what the court would regard as a reasonable amount of information in any given case.

What is the legal age for consent?

The legal age of consent for medical or surgical treatment is 16 years or over. In such cases, there is no legal requirement to obtain consent from a parent/guardian.

What is the legal position for consenting children less than 16 years of age?

A child under 16 may consent, if the health professional feel that they are capable of understanding the nature and possible consequences of the medical/surgical procedure (Fraser ruling). The same holds true for contraceptive advice. However, the value of parenteral support should be discussed. The health professional should also consider the best interests of the child. There is no lower age limit below which a health professional cannot give medical/surgical treatment.

Where a child is too young or is not considered to be Fraser ruling competent, then the parents must be involved in the consent process. One parent is usually sufficient. However for sensitive issues where opinions may vary, such as circumcision, consent should be sought from both parents.

What is the legal position of a child or parenteral refusal to treatment?

A Fraser competent child cannot refuse a treatment thought to be in his/her best interests in England and Wales. Parenteral consent will be needed.

It may be unlawful to proceed with treatment of a child considered not to be Fraser ruling competent in the face of parenteral refusal. Remember the recent MMR controversy.

How would you ensure quality in your unit?

How would you ensure that your team is up to scratch?

Use the seven pillars of clinical governance to answer the question. So I would ensure quality in my unit by:

- Education and Training – Supporting continuing professional development of all staff to ensure that they are competent and up to date.

- Clinical audit – Regular audits to review and improve clinical performance.

- Clinical effectiveness – by use of evidence based medicine.

- Risk management – by ensuring robust systems are in place to understand, monitor and minimise the risks to patients, staff and the organisation and to learn from mistakes and past experience.

- Research and development - by supporting research as well as ensuring that the research is implemented in patient care with a minimal time lag.

- Openness - promoting a blame free culture to ensure mistakes are reported/discussed and lessons learnt.

- Patient and public involvement -   to ensure services are responsive to the needs and aspirations of the patient and the public.

I also encourage reflective practice, provide regular feedback and ensure participation of all team members in decision making.

How do you think a blame-free culture can be brought about in the NHS?

In aviation, airline staff are rewarded for reporting mistakes and failures. More importantly, they can be disciplined, if not sacked, for not reporting them. That says something about the culture and the pride which airline workers have in their safety record.

What will it take to instill that sense of pride about safety in NHS staff? We recognise that there may be many barriers to creating a 'blame free' culture. But we feel that many of those barriers are psychological, the fear of what someone else could do to you. Part of that process is helping NHS staff realise they don't need to feel threatened or feel guilty about reporting after they've done so. We've got to look after the staff and recognise the traumas that many staff experience from being involved in adverse events.

What is the trust risk register?

A trust risk register is a management tool that enables an organisation to understand its comprehensive risk profile. It is simply a log of risks of all kinds that threaten an organisation's success in achieving its declared aims and objectives. It is a dynamic living document, which is populated through the organisation's risk assessment and evaluation process.

No	Source	Date in register	Description	Lead officer	Rating	Action summary	Completion date
1	Trust board	2/10/2009	Failure to meet waiting list targets	Chief executive	Extreme	Increase theatre capacity	Jan 2008
2	All wards	11/06/2010	Obsolete bed stock	Director of nursing	High	Capital bid for replacement	5 year programme

Give me an example of an adverse clinical incident you were directly involved in and how you handled this.

Quote a clinical incident which you were involved in or are aware of:

Steps in dealing with an adverse clinical incident- Ensure the safety of the patient first, involve your seniors, thorough documentation of the incident, debrief or actions taken to prevent its recurrence in future.

The general principles of dealing with an adverse clinical incident:

- What went wrong and why?
- What are the systems failures?
- What is the individual contribution in that failure?
- What lessons can be learnt?

You think a surgical emergency in the theatres has been mismanaged. What do you do?

- Number 1 priority – patient safety
- Discuss with surgeon and call for help
- Inform theatre sister and consultant in charge
- Careful documentation – critical incident – afterwards debrief

A nurse gives a substantially large dose of opioid to a patient? What do you do?

- Number 1 priority- check patient safety
- Careful documentation of the incident in the notes. Fill in an ireport form.
- Explain the incident to the patient
- Inform the charge nurse. Discuss the likely reasons- mistake, training deficits etc.
- Possible action - retraining, extra supervision

How would you ensure local delivery of national standards?

On an individual or team level, I will ensure delivery of national standards by:

- Regular audits to review and improve my practice
- Evidence based medicine by appropriately using NICE and other national guidelines
- Ensuring optimum resource allocation to ensure delivery of national standards
- Regular liaison with departmental heads and CCG
- Having procedures in place to remedy situations where practice is not in line with national standards
- Ensuring the team is aware of the goals and their roles in achieving it

What monitoring of standards would you undertake in your service?

- Audit compliance with clinical national standards, e.g. compliance with NICE guidance or guidance from relevant medical society
- Compliance with NSF
- Robust clinical governance mechanisms are in place
- Robust Risk management strategy
- Patient and public involvement in clinical services.

Caldicott 2

The original Caldicott Report (published in 1997) established six principles for NHS bodies to adhere to in order to protect patient information and confidentiality. The first review was exclusively concerned with the non-clinical use of patient data. The six principles resulting from the review aimed to preserve patients' trust by restricting the flow of information out of patient records.

The Future Forum report 2012 raised concerns regarding the current information governance arrangements. One of the principle concerns was that current information governance arrangements do not get the best balance between protecting patient confidentiality and sharing information to ensure high quality care. **There is reluctance to share clinical information by NHS staff due to fear of falling foul of current information governance arrangements**. This impacts clinical care. The Future Forum also noted that "it's the patient's data", not "the system's data". This was in response to frustration individuals faced when trying to access personal data held about them, their children or families caused by different organisations maintaining different policies on access.

The government thus commissioned Dame Fiona Caldicott to conduct a further Information Governance Review which was published in April 2013. The **review (Caldicott 2)** reinforces the six original Caldicott principles and makes one additional principle.

The added principle is: "The duty to share information can be as important as the duty to protect patient confidentiality".

The review highlights that that when it comes to sharing information, a culture of anxiety permeates health and social care organisations from the board room to front line staff. This anxiety results from instructions issued by managers to protect their organisation from fines for breaching data protection laws. This leads to a risk averse approach to information sharing even when it impacts patient care.

Accordingly, Recommendation 2 of the review specifically states that:

"for the purposes of direct care, relevant personal confidential data should be shared among the registered and regulated health and social care professionals who have a legitimate relationship with the individual."

The revised list of Caldicott principles therefore reads:

1. Justify the purpose(s)

Every proposed use or transfer of personal confidential data within or from an organisation should be clearly defined, scrutinised and documented, with continuing uses regularly reviewed, by an appropriate guardian.

2. Don't use personal confidential data unless it is absolutely necessary

Personal confidential data items should not be included unless it is essential for the specified purpose(s) of that flow. The need for patients to be identified should be considered at each stage of satisfying the purpose(s).

3. Use the minimum necessary personal confidential data

Where use of personal confidential data is considered to be essential, the inclusion of each individual item of data should be considered and justified so that the minimum amount of personal confidential data is transferred or accessible as is necessary for a given function to be carried out.

4. Access to personal confidential data should be on a strict need-to-know basis

Only those individuals who need access to personal confidential data should have access to it, and they should only have access to the data items that they need to see. This may mean introducing access controls or splitting data flows where one data flow is used for several purposes.

5. Everyone with access to personal confidential data should be aware of their responsibilities

Action should be taken to ensure that those handling personal confidential data — both clinical and non-clinical staff — are made fully aware of their responsibilities and obligations to respect patient confidentiality.

6. Comply with the law

Every use of personal confidential data must be lawful. Someone in each organisation handling personal confidential data should be responsible for ensuring that the organisation complies with legal requirements.

7. The duty to share information can be as important as the duty to protect patient confidentiality.

Health and social care professionals should have the confidence to share information in the best interests of their patients within the framework set out by these principles. They should be supported by the policies of their employers, regulators and professional bodies.

These principles should underpin information governance across the health and social care services.

The DoH has accepted all the recommendations of the review in Sep 2013.

Ref- Caldicott 2 report

http://tinyurl.com/cuye9b2

What is information governance (IG)?

Information Governance is the way by which the NHS handles all organisational information – in particular the personal and sensitive information of patients and employees. It allows organisations and individuals to ensure that personal information is dealt with legally, securely, efficiently and effectively, in order to deliver the best possible care.

It provides a framework to bringing together the requirements, standards and best practice that apply to the handling of information. It has four fundamental aims:

- To support the provision of high quality care by promoting the effective & appropriate use of information;
- To encourage responsible staff to work closely together, preventing duplication of effort and enabling more efficient use of resources;
- To develop support arrangements and provide staff with appropriate tools and support to enable them to discharge their responsibilities to consistently high standards;
- To enable organisations to understand their own performance and manage improvement in a systematic and effective way.

IG has come about because of concerns about public sector data protection. A board-level Senior Information Risk Owner (SIRO) is required in each organisation for IG.

Ref: http://tinyurl.com/czjl4v4

How do SIRO and Caldicott Guardian differ?

SIRO and Caldicott Guardian should work together. However,

SIRO
- Is accountable

- Fosters a culture for protecting and using data
- Provides a focal point for managing information risks and incidents
- Is concerned with the management of all information assets

The Caldicott Guardian

- Is advisory
- Is the conscience of the organisation
- Provides a focal point for patient confidentiality & information sharing issues
- Is concerned with the management of patient information

Ref: http://tinyurl.com/4vxs2v

Care Quality Commission

What is the role of Care Quality Commission (CQC)?

CQC is the independent regulator of health (NHS & private) and adult social care services in England, whether they're provided by the NHS, local authorities, private companies or voluntary organisations. They also protect the rights of people detained under the Mental Health Act.

CQC's job is to make sure that care provided by hospitals, dentists, ambulances, care homes and services in people's own homes and elsewhere meets national standards of quality and safety.

Under the Health and Social care Act 2012, the CQC will continue to act as the quality inspectorate across health and social care. The CQC's remit is distinct from Monitor in that its focus will be on quality and Monitor will focus on economic regulation. Under the new joint licensing regime (CQC and monitor), the CQC will be responsible for licensing NHS and adult social care providers against essential safety and quality requirements.

What are the national standards?

The national standards cover all aspects of care, including:

- Patients should expect to be respected, involved in your care and support, and told what's happening at every stage.
- Patients should expect care, treatment and support that meet their needs.
- Patients should expect to be safe.
- Patients should expect to be cared for by staff with the right skills to do their jobs properly.
- Patients should expect the care provider to routinely check the quality of their services.

CQC register care services that meet the standards, inspect them to check that they continue to do so, and take action when they don't.

How do CQC perform its function?

- CQC register health and adult social care services across England and inspect them to check whether or not standards are being met.

- Monitoring and inspection of all health and adult social care services. CQC inspections take place regularly and at any time in response to concerns. They are almost always unannounced. In between inspections CQC continually monitor all the information they hold about a service. The information comes from their inspections, the public, care staff, care services and from other organisations.

- Reporting the outcomes of their work on their website so that people who use services have information about the quality of their local health and adult social care services.

How do CQC enforce standards?

CQC have a range of powers to enforce standards. CQC can instruct care managers to produce a plan of action to make improvements or they can do the following:

- Issue a warning notice, asking for improvements within a short period of time.
- Restrict the services that the care provider can offer.
- Restrict admissions to the service.
- Issue a fixed penalty notice.
- Suspend the care provider's registration.
- Cancel the care provider's registration.
- Prosecute the care provider.

CQC work with local authorities, regulators and agencies, and sometimes the police, to make sure the necessary action is taken.

New inspection regime: Sep 2013

A new inspection regime was initiated by CQC in September 2013 because of concerns raised about the quality of its inspection in light of Mid Staffordshire scandal and other hospital inspections where its processes had missed significant problems. The new inspection regime is a specialist, thorough inspection with greater involvement of staff and the public. A new Chief Inspector of Hospitals was appointed by CQC to improve quality and drive up standards of care via CQC inspections.

Key changes in the new inspection regime:

- Larger and more specialised team: Bigger teams, headed by a senior clinician, and including both senior and junior clinicians and trained members of the public, would be used to inspect hospitals.

  The previous inspection regime was limited to four or five people, often not specialists in care.

- Inspection will focus on the 'whole patient experience': Each inspection will cover eight key services areas: A&E; medical care; surgery; critical care; maternity; paediatrics; end-of-life care and outpatients. The inspections will be a mixture of announced and unannounced visits and they will include inspections in the evenings and at weekends.

  That contrasts with the previous inspections which were grouped around essential standards so hospitals would find themselves inspected for issues such as nutrition and infection control rather than the entire system.

- In-depth inspection: Inspectors will now spend at least two days looking at the whole hospital.

- Greater public involvement: On the evening of the first day of each inspection there will be a 'listening event' where local people can tell members of the inspection panel their views of the hospital's care.

All 161 acute hospitals will be inspected under the new comprehensive inspection programme by 2015. There will continue to be additional inspections for trusts where concerns have been raised. The trusts that have been inspected will receive an overall rating—outstanding, good, requires improvement, or inadequate. Trusts will also be rated on the five domains- safety, effectiveness, caring, responsiveness, and leadership.

Complaints procedure

- Patient can contact Patient Advice Liaison Service i.e. PALS for help and advice or any complaints. Every hospital has PALS.

- Independent help and advice is also available from local Independent Complaints Advocacy Service (ICAS). ICAS is a national service that supports people who wish to make a complaint about their NHS care or treatment. They have a statutory role to support patients and carers who wish to make a complaint about their NHS treatment or care.

- If resolution is still not possible, ultimately, the patient can take their complaint to the Health Service Ombudsman. However you cannot do this simply because you are unsatisfied with the outcome; you must be able to provide reasons for their continued dissatisfaction and demonstrate that you are suffering continuing hardship or injustice, or that there is a reasonable prospect of achieving a worthwhile outcome. The ombudsman can then make a decision on whether or not the practice needs to carry out further investigations. The ombudsman can also decide to take on the complaint fully.

- In addition, a local HealthWatch would provide independent support to the patients.

The government (in its response to the Francis report) wants to see every trust make clear to every patient from their first encounter with the hospital:

- how they can complain to the hospital when things go wrong

- who they can turn to for independent local support if they want it, and where to contact them

- that they have the right to go to the Ombudsman if they remain dissatisfied, and how to contact them; and

- details of how to contact their local HealthWatch.

A poster (setting out how to complain about hospital, how to seek support from their local HealthWatch and how to refer their complaint to the Ombudsman) in every ward and clinical setting would be a simple means of achieving this. The government is liaising with HealthWatch England, the Care Quality Commission and NHS England as to the best means of ensuring this becomes standard practice in all NHS hospitals in England. The government also expects that local HealthWatch, as the patient and public champion for health and care services, should be as strong and effective as possible so that it can speak up for patients and provide independent support on complaints.

The Department of Health wants to see patient advice and liaison services (PALS) well-sign posted, funded and staffed in every hospital so patients can go and share a concern with someone else in the hospital if they do not feel confident talking to their nurse or doctor on the ward.

As part of its new inspection regime, the Care Quality Commission will be including complaints handling in its assessment of trust performance which includes how trusts have learnt from complaints.

Ref: http://tinyurl.com/c783f3t

http://tinyurl.com/l88ku85

Commissioning for Quality & Innovation (CQUIN)

What is CQUIN payment framework?

The CQUIN payment framework makes a proportion of provider income conditional on the achievement of ambitious quality improvement goals and innovations agreed between the commissioner and provider, with active clinical engagement. The CQUIN framework is intended to encourage a culture of continuous quality improvement and innovation in all providers. The framework was launched in April 2009.

What are local CQUIN schemes?

A CQUIN scheme is the agreed package of goals and indicators, which in total, when achieved, enables the provider to earn its full CQUIN payment (2.5% of contract value in 2013/14).

What is a CQUIN goal?

A CQUIN goal is a description of the intended objective which is being incentivised by the CQUIN scheme. The goals should cover the three domains of quality – safety, effectiveness and patient experience – and innovation. In 2013/14, national CQUIN schemes for acute providers include Friends and Family test, improvements in NHS safety thermometer, improving dementia care and 95% patients being risk assessed for VTE.

Each CQUIN goal must be measurable, using a defined indicator.

Does the CQUIN framework apply to independent sector providers?

The CQUIN payment framework applies to all services covered by national standard contracts in 2013/14. This means it will cover acute, community, mental health, ambulance and specialised services, as well as independent sector providers and foundation trusts on national standard contracts.

How does the CQUIN framework relate to pre-existing quality schemes and quality improvement plans?

The CQUIN payment framework is not intended to replace pre-existing local quality initiatives. The CQUIN framework is only one driver for quality and quality should be at the heart of all commissioning decisions and strategic plans.

Ref: http://tinyurl.com/pnvo8pc

Consultant Outcomes Publication

Consultant Outcomes Publication (COP) is an NHS England initiative, managed by HQIP (Healthcare Quality Improvement Partnership), to publish quality measures at the level of individual consultant doctor using National Clinical Audit and administrative data. The data is published on the NHS choices website (http://tinyurl.com/qykr277).

The information published so far includes how many times each participating consultant has performed certain procedures and what their mortality rate is for those procedures. The data shows where the clinical outcomes for each consultant sit against the national average. The data is risk adjusted to ensure outcomes are calculated as if all consultants operated on the 'average' patient.

COP began in 2013. The medical specialties included in COP 2014 includes Adult Cardiac surgery, Bariatric surgery, Colorectal surgery, Head and neck surgery, Interventional Cardiology, Thyroid and Endocrine surgery, Orthopaedic surgery, Upper GI surgery, Urological surgery, Vascular surgery, Lung cancer, Urogynaecology and Neurosurgery.

The aim of COP is to drive up the quality of care in the NHS and improve transparency.

Prof Sir Bruce Keogh, National Medical Director of NHS England, said: 'We know from our experience with heart surgery that putting this information into the public domain can help drive up standards. That means more patients surviving operations and there is no greater prize than that'.

The reporting of the data was led by Prof Ben Bridgewater from the Healthcare Quality Improvement Partnership (HQIP). Prof Bridgewater is a practising heart surgeon who leads the successful cardiac consultant-level reporting which paved the way for this work.

Prof Bridgewater said: 'Ultimately there is one patient and one responsible consultant. This means the public can now know about the care given by each doctor and be reassured an early warning system is in place to identify and deal with any problems

Due to data protection legislation, consultants had to agree to have results from their operations published and around 98% have. The names of those consultants who have not agreed to have data published and the trusts they work in can be seen on the NHS choices website.

Some surgeons object to the principle of attributing surgical results to an individual when those results are dependent on effective teamwork between surgeon, anaesthetist, theatre and ward nurses and physiotherapists. Prof Sir Bruce Keogh counters that the patient enters the agreement for surgery with the surgeon and someone has to be accountable for the team's outcomes.

What will the NHS do where consultants have high mortality rates?

Any hospital or consultant identified as an outlier will be investigated and have action taken to improve data quality and/or patient care.

Duty of Candour

The General Medical Council (GMC) (with eight UK professional healthcare regulators) has underlined its commitment to a professional duty of candour for doctors in a statement issued in Oct 2014

Health professionals must be open and honest with patients when things go wrong. This is also known as 'the duty of candour'.

Every healthcare professional must be open and honest with patients when something goes wrong with their treatment or care which causes, or has the potential to cause, harm or distress.

This means that healthcare professionals must:

• tell the patient (or, where appropriate, the patient's advocate, carer or family) when something has gone wrong

• apologise to the patient (or, where appropriate, the patient's advocate, carer or family);

• offer an appropriate remedy or support to put matters right (if possible); and

• explain fully to the patient (or, where appropriate, the patient's advocate, carer or family) the short and long term effects of what has happened.

Healthcare professionals must also be open and honest with their colleagues, employers and relevant organisations, and take part in reviews and investigations when requested. Health and care professionals must also be open and honest with their regulators, raising concerns where appropriate. They must support and encourage each other to be open and honest and not stop someone from raising concerns.

'The awful reality that emerged from Mid Staffs and indeed other inquiries was that doctors knew about GMC guidance but were not empowered by it. They felt it was acceptable to 'walk by the other side of the ward' knowing that there was unsafe and unacceptable practice going on. We must all do what we can to make sure that does not happen again. The statement above is an important milestone and makes it clear that the professional duty of candour sits with every healthcare professional, regardless of their field of practice.'

The government in November 2014 introduced a further duty of candour on secondary care organisations registered with CQC - one required, and enforceable, by law.

This new statutory duty of candour will apply to all other care providers registered with CQC from 1 April 2015. The key principles are:

1. Care organisations have a general duty to act in an open and transparent way in relation to care provided to patients. This means that an open and honest culture must exist throughout an organisation.

2. The statutory duty **applies to organisations, not individuals,** though it is clear from CQC guidance that it is expected that an organisation's staff cooperate with it to ensure the obligation is met.

3. As soon as is reasonably practicable after a notifiable patient safety incident occurs, the organisation must tell the patient (or their representative) about it in person.

4. The organisation has to give the patient a full explanation of what is known at the time, including what further enquiries will be carried out. Organisations must also provide an apology and keep a written record of the notification to the patient.

5. A notifiable patient safety incident has a specific statutory meaning: it applies to incidents where a patient suffered (or could have suffered) unintended harm that results in death, severe harm, moderate harm or prolonged psychological harm.

6. There is a statutory duty to provide reasonable support to the patient.

7. Once the patient has been told in person about the notifiable patient safety incident, the organisation must provide the patient with a written note of the discussion, and copies of correspondence must be kept.

Doctors are most likely to be the organisation's representative under the statutory duty. It is important that you cooperate with your organisation's policies and procedures, including the requirement to alert the organisation when a notifiable patient safety incident occurs.

An area of difficulty may be deciding whether an incident reaches the threshold for notification under the statutory duty. This may be confusing, as the threshold is low for the doctor's ethical duty (any harm or distress caused to the patient) while the thresholds for the contractual and statutory duties are higher and slightly different (at least moderate harm).

Ref: http://tinyurl.com/longfbo

Euthanasia and Assisted Suicide

Euthanasia and assisted suicide is against the law in the UK. BMA and Royal Colleges are against all forms of assisted suicide.

There are arguments both for and against euthanasia and assisted suicide. Some of the main arguments are outlined below.

Arguments for euthanasia and assisted suicide

There are two main types of argument used to support the practices of euthanasia and assisted suicide. They are the:

Ethical argument – that people should have freedom of choice, including the right to control their own body and life, and that the state should not create laws that prevent people being able to choose when and how they die. The ethical argument suggests that life should only continue as long as a person feels their life is worth living.

Pragmatic argument – that euthanasia, particularly passive euthanasia (like many of the practices used in end of life care), is allegedly already a widespread practice, just not one that people are willing to admit to, so it is better to regulate euthanasia properly.

Arguments against euthanasia and assisted suicide

Religious argument – that these practices can never be justified for religious reasons; for example, many people believe that only God has the right to end a human life.

'Slippery slope' argument – this is based on the concern that legalising euthanasia could lead to significant unintended changes in our healthcare system and society at large that we would later come to regret.

Medical ethics argument – that asking doctors, nurses or any other healthcare professional to carry out euthanasia or assist in a suicide would be a violation of fundamental medical ethics.

Alternative argument – that there is no reason for a person to suffer either mentally or physically because effective end of life treatments are available; therefore, euthanasia is not a valid treatment option, but represents a failure on the part of the doctor involved in a person's care.

Ref: http://tinyurl.com/7mbce5q

Assisted dying around the world

The Netherlands: Voluntary euthanasia is legal for the terminally ill and in cases of hopeless and unbearable suffering that cannot be alleviated.

Germany: Assisting somebody by providing the means for their suicide is within the law.

Norway: Causing or contributing to someone's death can be punished by up to eight years in prison. A carer who helps to end the life of a consenting person who is terminally ill, on compassionate grounds, may receive a reduced sentence.

Switzerland: Assisted dying is legal. Volunteers have to collect the lethal medication from a pharmacy and take it to the person who wants to die.

Foundation Trusts (FTs)

What is a Foundation Trust?

They are autonomous organisations, free from central government control. They establish strong connections with their local communities through local people becoming members and governors. This enables foundation hospitals to design their healthcare services around local needs and priorities. Foundation hospitals are firmly part of the NHS, providing healthcare according to core NHS principles; free care, based on need and not ability to pay.

They are able to decide for themselves what capital investment is needed in order to improve their services and are free to retain any surpluses they generate and to borrow in order to support this investment.

History: FTs were established by the Health and Social Care Act 2003 to allow NHS trusts more independence from central government control. Monitor was established in 2004, as the 'Independent Regulator of FTs'. Independent of government and directly accountable to Parliament, its remit was to undertake the authorisation and regulation of FTs. There were concerns when FTs were introduced, that the new financial arrangements could incentivise prioritisation of private patients over NHS services. In response to this the NHS Act 2006 introduced a private patient income (PPI) cap, which fixed the level of private patient income earned by an FT to its 2002-03 level. Today the total amount of income earned by FTs from private patients varies considerably from nothing to almost 30 percent of income.

Under the Health and Social care Act 2012, the Private Patient Income cap is set at 49 percent. However, an increase in the proportion of an FT's private income of more than 5 percent would need majority approval by its governors and every FT must set out how its non-NHS income has benefited NHS services in its annual report.

What are the pros and cons of FTs?

Benefits to Patients

- NHS FTs will be able to improve relevant care for their patients because they have been set free from central government control.
- Crucially, NHS Foundation Trusts will have the freedom to decide locally the capital investment needed in order to improve their services and increase their capacity. They will be able to borrow in order to support this investment without needing to seek external approval.
- A membership body comprising local staff & service users will elect governors. This form of public ownership & accountability will ensure that hospital services more accurately reflect the needs & expectations of local people.

Concerns regarding FTs

FTs will lead to greater inequalities between hospitals by:

- FTs will be able to keep all operating surpluses & asset sale proceeds themselves, whereas under the current system surpluses go to a central NHS funding pool from where they are redistributed to wherever the need in the NHS is the greatest.
- FTs will draw scarce staff away from non foundation trusts. FTs will have greater powers to raise private funds & set wage levels & will therefore be able to exercise additional flexibility on pay, leading to these hospitals drawing scarce staff away from trusts that do not have foundation status.
- Because private sector borrowing by foundation hospitals will be counted against government expenditure limits, this may leave less money for non-foundation hospitals - robbing Peter to pay Paul.

What is the governance structure of foundation hospitals?

Foundation hospitals have a new form of governance structure that involves members, a Board of Governors, and a Management Board.

The members are drawn from patients, staff and local people. Once a year, the members of the foundation hospital elect representatives to its Board of Governors. This will have the responsibility for approving the annual report and accounts, for setting the foundation hospital's strategic direction, and for ensuring that it does not breach the terms of its license.

A majority of the members of the Board of Governors must be members of the public, and it must also include at least one staff member and one representative each from the hospital's main commissioning CCG and any universities responsible for undergraduate training at the hospital.

As well as the Board of Governors, each foundation hospital will have a Management Board. This will have a duty to consult the Board of Governors concerning the development of the hospital's forward plans and regarding any significant changes to the business plan.

The Management Board will be chaired by the chair of the Board of Governors and at least a third of the places excluding the chair must be filled by non-executive directors elected by the Board of Governors. In addition the Management Board must include a Chief Executive, who will be appointed by the chair and nonexecutive directors of the management board, a medical director and a finance director.

What is the role of MONITOR?

Monitor was set up as an independent corporate body to authorise and regulate NHS foundation trusts to ensure that they are well managed and financially strong.

Under the Health and Social Care Act 2012, Monitor's previous role of oversight of foundation trusts will continue and the Care Quality Commission (CQC) will continue to act as the quality inspectorate across health and social care. Monitor will focus on financial regulation while CQC will focus on quality.

Discuss how foundation hospitals have performed since introduced in 2004.

A health committee reported into foundation trusts in 2008.

- It is clear that the majority of FTs are high performers in terms of finance and quality as measured by CQC ratings. However, these were high performing organisations prior to becoming FTs, and so it is difficult to ascribe this high performance to FT status per se.
- A major advantage of FT status is the autonomy it gives trusts. FTs do not appear to have yet exploited the full potential of their autonomy, however they are free to make decisions more quickly, & that this is making a 'tangible' difference to the dynamic of these organisations.
- The governance of FTs (i.e. a membership body comprising local staff and service users, who elect governors) was set up to promote better local engagement and thus hospital services to more accurately reflect the needs and expectations of local people. However, this governance structure has been slow to deliver benefits and there is lack of robust evidence of their effectiveness.
- Before their establishment a number of fears were voiced about the impact FTs might have on wider health communities. There is little evidence that FTs have poached staff from other trusts. FTs may be generating tensions and resentment in some areas, however it is felt these tensions exist between high performing and less well performing trusts regardless of their status because of the system of payment by results.

- A major concern at the inception of FTs was that they, together with payment by results, would strengthen the acute sector to the detriment of primary care services. This seems to be the case, although it is probably more because of the introduction of payment by results than the introduction of FTs.

So in summary, FTs have some proven strengths, but much is unknown. In general, robust evidence is lacking. It is not clear whether their high performance in terms of finance and quality is the result of their changed status, or simply a continuation of long term trends, since the best trusts have become FTs. Key aims of FTs were the promotion of innovation and greater public involvement, but, again, there is a lack of objective evidence about what improvements, if any, FTs have produced.

Difference between Foundation and NHS Trusts		
	Foundation Trust	NHS Trust
Government regulation	Not directed by government	Directed by government
Financial regulation	Monitor	Trust Development Authority
Quality	CQC	CQC
Finance	Free to make their own financial decisions.	Financially accountable to government

Health and Social Care Act 2012

The Health and Social Care Act 2012 introduced radical changes to the way that the NHS in England is organised. The legislative changes from the act came into being on 1 April 2013 and include:

A. A move to clinically-led commissioning. Planning and purchasing healthcare services for local populations had previously been performed by England's 152 primary care trusts (PCTs). The act replaced the PCTs with 211 clinical commissioning groups (CCGs), led by clinicians. CCGs now control the majority of the NHS budget, with highly specialist services and primary care being commissioned by NHS England.

B. An increase in patient involvement in the NHS. The act established independent consumer champion organisations locally (HealthWatch) and nationally (HealthWatch England) to drive patient and public involvement across health and social care in England. The HealthWatch network has significant statutory powers to ensure the voice of the consumer is strengthened and heard by those who commission, deliver and regulate health and care services.

C. A renewed focus on the importance of public health. The act provided the legislation to create Public Health England (PHE), an executive agency of the Department of Health. PHE's aim is to protect and improve the nation's health and to address health inequalities.

D. A streamlining of 'arms-length' bodies. The act conferred additional responsibility on the National Institute for Health and Care Excellence (NICE – formerly the National Institute for Clinical Excellence) to develop guidance and set quality standards for social care. The Health and Social Care Information Centre (HSCIC) was also tasked with responsibility for collecting, analysing and presenting national health and social care data.

E. Allowing healthcare market competition in the best interest of patients. The act aimed to allow fair competition for NHS funding to independent, charity and third-sector healthcare providers, in order to give greater choice and control to patients in choosing their care. To protect the interests of patients under these new arrangements, Monitor was established as the sector economic regulator for health services in England. Monitor issues licences to NHS-funded providers, has responsibility for national pricing and tariff (in conjunction with NHS England) and helps commissioners ensure that local services continue if a provider is unable to continue providing services. The Care Quality Commission (CQC) will continue to act as the quality inspectorate across health and social care. The government says that the CQC's remit is distinct from Monitor in that its focus will be on quality.

F. Health Education England (HEE) was established as a Special Health Authority in June 2012 to promote high quality education and training as well as authorising and supporting Local Education and Training Boards (LETBs). It is expected that the education and training functions of SHAs and postgraduate deaneries will be undertaken by LETBs. These LETBs will provide a forum for workforce development to support research and innovation, coordinating workforce planning activity, and commissioning education and training locally.

Health Education England (HEE)

HEE was established as a Special Health Authority in June 2012. It assumed full operational responsibilities from April 2013.

HEE provide leadership for the new education and training system. The driving principle for reform of the education and training system is to improve care and outcomes for patients and HEE exists for one reason alone – to help ensure delivery of the highest quality healthcare to England's population, through the people we recruit, educate, train and develop.

HEE will ensure that the shape and skills of the future health and public health workforce evolve to sustain high quality outcomes for patients in the face of demographic and technological change. HEE will ensure that the workforce has the right skills, behaviours and training, and is available in the right numbers, to support the delivery of excellent healthcare and drive improvements. HEE will support healthcare providers and clinicians to take greater responsibility for planning and commissioning education and training through the thirteen Local Education and Training Boards (LETBs), which are statutory committees of HEE.

HEE holds a budget of £4.9 billion for multi-professional education and training, which it distributes to LETBs.

The key national functions of the organisation include:

- Providing national leadership for planning and developing the whole healthcare and public health workforce
- Authorising & supporting development of Local Education & Training Boards & holding them to account
- Promoting high quality education and training which is responsive to the changing needs of patients and communities and delivered to standards set by regulators

- Allocating and accounting for NHS education and training resources – ensuring transparency, fairness and efficiency in investments made across England.
- Ensuring security of supply of the professionally qualified clinical workforce
- Assisting the spread of innovation across the NHS in order to improve quality of care
- Delivering against the national Education Outcomes Framework to ensure the allocation of education and training resources is linked to quantifiable improvements.

Ref: www.hee.nhs.uk.

Healthcare Quality Improvement Partnership (HQIP)

Major efforts have been put in place to improve and address quality in health and social care services in the UK. HQIP was one such initiative to drive up quality. It was established in 2008 with the responsibility to reinvigorate national clinical audit and promote quality in healthcare.

HQIP

- It is a body owned and led by professional bodies (academy of Royal Colleges, Royal College of Nursing etc) and patient/service user interest groups. It's a unique alliance between patient and professional to lead QI programmes that are rooted in professional practice and which matter to the patients

- HQIP is a charity and company limited by guarantee.

- HQIP sees clinical audit as an essential tool in QI. However, it will promote any improvement methodology that has proven effectiveness and appropriateness.

- HQIP fosters both national and local level partnerships between clinicians, clinical teams, managers and patients to do effective QI work.

- It supports local staff, fosters active dissemination of information and implements quality improvement initiatives to ensure quality measurement is the engine which drives improvement.

In 2011 HQIP extended its work to promote quality improvement within social care.

Ref: www.hqip.org.uk

Independent sector treatment centers (ISTC)

What are Independent sector treatment centres?

ISTC are private companies set up to reduce waiting lists for patients awaiting elective procedures. ISTC are run either by the NHS or commissioned by CCGs from independent sector providers. ISTC thus provides NHS funded care in a private setting. They operate under the banner of NHS and use NHS branding and logo. ISTC can only employ doctors who are registered with the GMC

What are the pros and cons of Independent sector treatment centres?

Benefits of ISTCs

- Provides extra clinical capacity to deliver swift access to treatment for NHS patients

- Provides diversity and choice in clinical services for NHS patients

- Stimulate innovative models of service delivery

- Drive up productivity

Issues caused by ISTCs

- There are doubts whether ISTCs provide value for money.

- ISTCs are given 5 year contracts that guarantee payment regardless of whether they carry out the specified number of operations or not. This results in an increased cost to the NHS.

- ISTCs deal with simple elective procedures and therefore take away training opportunities from the NHS.

- Only provide profitable services (for example hernias and varicose veins) by "cherry-picking" suitable patients - that is, those with minimal co-morbidity.

Integrated Care

What is integrated care?

Integrated care means care which is organised around the needs of individual patients. Integrated care is not about structures, organisations or pathways – it is about better outcomes for service users.

There is now a clear consensus that successful integrated care is primarily about patient experience, although all dimensions of quality and cost-effectiveness are relevant.

In order to show why a smooth journey through the NHS is desirable, it is important to understand what a non-integrated system looks like and the experience that service users face when navigating through a fragmented system. For example, a diabetic patient may need services of the following: cardiology team, GP, practice nurse, renal team, diabetic team, diabetes specialist nurse, GP, practice nurse, vascular team, foot team, eye team, retinal screening team etc. This illustrates the complexity service users must navigate and, therefore, highlight the need of a more integrated system.

What are the problems with the current non-integrated care system?

- Lack of 'ownership' for the patient and her problems, so that information gets lost as the patient navigates the system
- Lack of involvement by the user/patient in the management and strategy of care
- Poor communication with the user/patient as well as between health and social care providers
- Treating service users for one condition without recognising other needs or conditions, thereby undermining the overall effectiveness of treatment

- Lack of integration between health and social care- although decisions made in the social care setting affect the impact of health care treatment, and vice versa.

Is integrated care really needed?

The NHS is faced with the major challenges of using resources more efficiently and of meeting the needs of an ageing population in which chronic medical conditions are increasingly prevalent. The key task therefore is to implement a new model of care in which clinicians work together more closely to meet the needs of patients and to coordinate services and enable people with complex needs to live healthy, fulfilling, independent lives.

This model of integrated care would focus much more on preventing ill health, supporting self-care, enhancing primary care, providing care in people's homes and the community, and increasing co-ordination between primary care teams and specialists and between health and social care

Integrated care is felt to be the only way to make health system sustainable in the long term by transferring care out of expensive hospitals and nursing homes in the community or even in patients own home.

What are the aims of integrated care?

- Improved patient experience- by providing a seamless service reducing gaps and duplication in service
- Improved patient outcomes i.e. improved quality
- Improved cost effectiveness of care by improving system efficiency
- Better experience for medical staff

However it is worth remembering that more integrated care is not always the right answer to improving the patient's experience and system efficiency. Integrated care also carries some risks, such as that of reducing competition, and incentives to improve quality.

How would you develop integrated care?

- There are no general rules
- Benefits depend on the specific design and approach to integrated care based on local circumstances
- Learn from successful examples
- Take patients' views into account

Successful integration depends on which approach is used, how well it is implemented, and on features of the environment in which a provider is operating, including the financing system.

What are the barriers to integration?

- **Organisational boundaries**- between Primary and secondary care, social care (provided by LA) and health care- making it difficult for services to be properly coordinated. This leads to gaps and duplication of services. In a fully integrated system, patients' needs not organisational boundaries would decide how care is provided
- **Lack of shared record keeping** about a patient (GPs, hospital, nurses, social care workers) - hence patient and carers end up telling the same story to healthcare professionals.

There are other barriers like:

- Payment by results provides hospitals with some incentives to keep patients in hospital rather than treating them in the community (this is changing with the updated guidance on PbR).

- Service users choosing alternative providers: service users have freedom of choice regarding their elected place of care. However, this freedom can create deviations from the planned pathway of care and may cut across attempts to provide integrated care.

How can organisations enable integrated care?

Several measures could potentially mitigate these barriers. They may include:

- Personal budgets
- Making it easier for service users and carers to coordinate and navigate. This implies that every service user with long-term or complex needs has easy access to a "care coordinator", or federations of GP practices who can act as the coordinating point for all of their care.
- Information is a key enabler of integrated care. Care records should be electronic and accessible at the point of care throughout the whole care journey, regardless of sector or provider.

Can we do it?

NHS has spent the last decade or more in ensuring faster and easier access to GPs and consultants; reducing waiting times etc. and indeed vast progress has been made in these areas.

This was made to matter to the managers and clinical leaders- we need the same focus- the same urgency and importance being attached to integrated care for patients with long term conditions, frail older people- and only when it matters that integrated care will become a reality.

What can we do to facilitate integrated care?

- Clear policy directive as to why integrated care is important; we spend too much in acute hospitals, care to individual patient is poor and if we don't change the system, it will never become a sustainable system

- Sustained support for people seeking to develop integrated care locally- support in the form of using data to analyse population need, project management etc

- Evaluation of integrated care initiatives so that we can learn

In summary, integrated care means that things are joined up, not fragmented into different parts with no one person informing all the relevant agencies about patients' needs and their conditions. Importantly, integrated care should also include the person's wishes and needs, too often at the moment the person is left out of the loop!

Guidance for Taking Responsibility: Accountable Clinicians & Informed Patients

The Francis Report made a number of recommendations on the need for there to be a named clinician who is accountable for a patient's care whilst they are in hospital. In addition, the Secretary of State for Health in England has supported the concept of having an accountable consultant and nurse with their "name over the bed".

The Academy of Medical Royal Colleges (the Academy) was asked by the Secretary of State for Health to see how this could be taken forward.

The AoMRC released the following guidance in Jun 2014.

- The Responsible Consultant/Clinician is an individual named consultant/clinician who has responsibility for the overall management, continuity and delivery of all care to a patient throughout their hospital stay. Wherever possible, the responsible consultant/clinician should remain the same for the duration of a patient's hospital stay. There may be occasions when it is clinically appropriate that the role is formally transferred to another consultant with the documented agreement of all parties.

- Whilst s/he may not be individually accountable for the delivery of every aspect of a patient's care the responsible consultant/clinician is the person to whom a patient or their relative/carer would ultimately address concerns about any aspect of care throughout their stay. This means they will take overall responsibility for ensuring that any clinical issues, reports of specialised tests or investigations, difficulties or complaints are addressed appropriately.

- The responsible consultant/clinician may seek advice, investigations, treatment or patient reviews from other clinicians. Individual clinicians and other staff providing specific elements of the care of a patient during their stay in hospital (e.g. diagnostics, nursing, therapy, or administration) should retain their full organisational and professional accountability for all their actions.

- If an individual with the duties of a responsible consultant/clinician ignores or fails to take that overall responsibility, they may be considered as acting in an unprofessional manner by their employer or professional regulator.

- However, aside from individual clinicians taking on the role of responsible consultant/clinician, this concept will only work effectively if it is supported and enabled by employers.

Allocating the role of responsible consultant/clinician

- It is expected that the responsible consultant/clinician role would be determined at the point a patient is admitted as an in-patient. It is not expected that an emergency care clinician would be the responsible consultant/clinician beyond the emergency department.
- For patients admitted initially to the Medical/Surgical/Intensive Care Admissions Unit, a responsible consultant/clinician would be allocated to each patient for the admission period. However, should a patient then be transferred to a ward the responsible consultant/clinician will be reassigned unless they are transferred to a ward with the same responsible consultant/clinician.
- Patients classed as outliers should remain under the care and responsibility of their named responsible consultant/clinician.

- When a patient with the same clinical condition requires readmission to hospital, they should where possible remain under the care of their previous responsible consultant/clinician unless this would cause undue delay to treatment.

Displaying Information

The information, however, should be displayed in an appropriate manner near the patient's bedside. There is obviously no single correct solution but it is suggested that the two principles to be followed should be:

— Display relevant information in the relevant place e.g. if visitors require the information, it needs to be at the bedside. If it is information only required by nursing or other clinical staff it may be more appropriate at the nursing station

— The information displayed should be clear and simple.

National Clinical Assessment Service (NCAS)

NCAS is an operating division of NHS Litigation Authority (NHS LA) since 2013. The NCAS promotes patient safety by providing confidential advice and support to the NHS in situations where the performance of doctors' dentists and pharmacists are giving cause for concern. Managers or practitioners themselves can contact NCAS for advice. There was concern that tackling problems with medical performance needed specialist skills which were not always available in individual NHS trusts. Before NCAS existed, concerns could be referred to the GMC even when not serious enough to justify regulatory action.

NCAS process

- If a concern comes to light, the employer or practitioner can contact NCAS for help. NCAS will to work with all parties to clarify the concerns, understand what is leading to them, and make recommendations on how they may be resolved.
- The support NCAS provides includes not only advice over the telephone, but also more detailed and ongoing support. Where the performance problem is sufficiently serious or repetitious, and attempts to resolve the problem at local level have failed, a practitioner may be asked to undergo an NCAS assessment. This comprises three main components: an occupational health assessment, a behavioural assessment and a clinical assessment (by a team of clinical and lay assessors). A report containing the findings, conclusions and recommendations is produced by the whole NCAS assessment team. NCAS

will then work with the practitioner and the referring body to agree an action plan to resolve the concerns.

- NCAS is an advisory body only. It does not function as a regulator.

Where a concern about a practitioner's performance arises and the employer or contractor feels they need help, approaches to three different organisations are often considered: the GMC, NCAS or the medical royal college covering the relevant clinical specialty. What then guides the approach taken is broadly as follows:

- If the concern, whether of performance, health or conduct, is so serious as to call into question the doctor or dentist's license to practice, the regulator's advice should be taken. This approach will therefore only be used in the most serious circumstances.
- On the other hand, if the concern is more broadly based about a whole clinical service rather than about one or more individuals within a team, or where the organisation is unsure whether the treatment of a specific group of patients has met accepted standards, the colleges are often contacted for advice.
- In all other circumstances – such as immediate concerns that might require exclusion or suspension – general concern about a practitioner's performance, conduct or competence, and in any situation where the local organisation is unsure how to proceed, NCAS should be contacted.

In any event, all of those organisations work closely together and have published memoranda of understanding outlining how they work together. Contact with any of them will enable a discussion of how a concern is best handled and which agencies should be involved.

Ref: www.ncas.nhs.uk

No decision about me, without me

The slogan, 'No decision about me, without me', was originally a demand formulated by the emerging patient movement. The current government adopted the 'no decision' slogan in its 2010 White Paper, Equity and Excellence: Liberating the NHS.

The government wants to place patients' needs, wishes and preferences at the heart of clinical decision-making.

Why is shared decision-making important?

- Shared decision-making is viewed as an ethical imperative by the professional regulatory bodies which expect clinicians to work in partnership with patients, informing and involving them whenever possible.

- There is also compelling evidence that patients who are active participants in managing their health and health care have better outcomes than patients who are passive recipients of care.

- Shared decision-making is also important for commissioners because it reduces unwarranted variation in clinical practice. Shared decision-making is the principal mechanism for ensuring that patients get the care they need and is the essential underpinning for truly patient-centred care delivery.

Making shared decision-making a reality: No decision about me, without me aims to clarify shared decision-making and what skills and resources are required to implement it and it also outlines what action is needed to make this vision a reality.

The principle of shared decision-making in the context of a clinical consultation is that it should:

- support patients to articulate their understanding of their condition and of what they hope treatment (or self-management support) will achieve

- inform patients about their condition, about the treatment or support options available, and about the benefits and risks of each

- ensure that patients and clinicians arrive at a decision based on mutual understanding of this information

- record and implement the decision reached.

The paper outlines the importance of communication skills and sets out how clinicians might approach consultations to arrive at shared decisions. It also suggests that tools that help patients in making decisions are just as important as guidelines for clinicians. In fact, the government officially launched a set of innovative online tools that can help patients make informed decisions about their healthcare.

Further reading: http://tinyurl.com/6av4dbk

Online tools: http://sdm.rightcare.nhs.uk/pda/

National Confidential Enquiry into Patient Outcome and Death (NCEPOD)

NCEPOD review medical clinical practice and make recommendations to improve the quality and safety of patient care. They do this by undertaking confidential surveys covering many different aspects of medical care and making recommendations for clinicians and management to implement.

Primarily NCEPOD exists to alert clinicians and hospital management to practice which may not have been of the best quality and to recommend improvements. However, it does not audit individual clinician's performance and it therefore has no direct involvement in individual patient care and is not able to provide medical opinions or recommendations to specialists.

How is NCEPOD governed and funded?

NCEPOD is independent of the Department of Health and the professional associations. It is both a charity and a company limited by guarantee. NCEPOD has a board of directors that are referred to as the NCEPOD Trustees. In addition, there is a NCEPOD Steering Group. Members are nominated representatives of the various medical Royal Colleges and Associations and lay representation.

The work is commissioned by the Healthcare Quality Improvement Partnership on behalf of the funding bodies listed below and additionally funded by the independent sector hospitals. Funding bodies include;

- Department of Health, England
- Welsh Assembly Government
- Department of Health, Social Services and Public Safety, Northern Ireland
- States of Jersey
- States of Guernsey
- Isle of Man Government

How does NCEPOD select studies?

Each year, NCEPOD invites organisations or individuals to submit original study proposals for consideration as possible forthcoming studies. Proposals should be relevant to the current clinical environment and have the potential to contribute original work to the subject.

What happens if NCEPOD find a case that gives them cause for concern?

Cases that cause NCEPOD concern are referred back to the medical director of the trust concerned in order that appropriate action may be taken. Consultants involved with the case are also notified.

Is it mandatory to participate in the work of NCEPOD?

Yes it is. The DoH in their guidance on clinical governance state that trusts must take part.

Give an example of a NCEPOD report?

Quote a report relevant to your specialty (if there is one)

Examples of NCEPOD reports are scoping our practice, the acute problem, which operates when, the corners autopsy etc.

Ref: http://www.ncepod.org.uk

NHS constitution

The Constitution was developed as part of the NHS Next Stage Review led by Lord Darzi. The NHS Constitution was published on 21 January 2009 and applies to NHS services in England.

The Constitution brings together what staff, patients and public can expect from the NHS. It outlines the purpose, principles and values of the NHS, and highlights a number of rights, pledges and responsibilities for staff and patients.

All NHS bodies and private and third sector providers supplying NHS services will be required by law to take account of this Constitution in their decisions and actions. The Constitution will be renewed every 10 years, with the involvement of the public, patients and staff.

An updated NHS constitution was issued in March 2013. The NHS constitution has four parts: NHS principles, NHS values, patients and the public (rights and responsibilities) and staff (rights and responsibilities).

1. Principles that guide the NHS: Seven key principles guide the NHS in all it does

- The NHS provides a comprehensive service, available to all.

- Access to NHS services is based on clinical need, not an individual's ability to pay.

- The NHS aspires to the highest standards of excellence and professionalism.

- The NHS aspires to put patients at the heart of everything it does.

- The NHS works across organisational boundaries and in partnership with other organisations in the interest of patients, local communities and the wider population.

- The NHS is committed to providing best value for taxpayers' money and the most effective, fair and sustainable use of finite resources.

- The NHS is accountable to the public, communities and patients that it serves.

2. NHS values

- Working together for patients- Patients come first in everything we do.
- Respect and dignity- value every person, whether patient, their families or carers, or staff.
- Commitment to quality of care.
- Compassion- we ensure that compassion is central to the care we provide.
- Improving lives- We strive to improve health and wellbeing and people's experiences of the NHS.
- Everyone counts- We make sure nobody is excluded, discriminated against or left behind.

3a. Patients and Public- your rights and pledges

Access: The right to receive free (in most cases) equitable, timely access to treatment, and to expect your local NHS to meet the needs of the local community. The NHS also commits to provide convenient, easy access to services within the waiting times set out in the handbook to the NHS Constitution. The NHS also pledges to make the transition as smooth as possible when patients are referred between services

Quality of care and environment: The right to receive a professional standard of care, in a clean and fit for purpose environment, and for organisations to monitor and try to improve the standards of care that they provide, and to identify and share best practice with each other.

Nationally approved treatments and drug programmes: The right to receive drugs and treatments recommended by NICE if deemed clinically appropriate, and to receive vaccinations provided through national programs.

Respect consent and confidentiality: Service users should be treated with dignity and respect, expect privacy and confidentiality and be given information about treatments that are being given. This includes sharing any letters that are sent between clinicians. Patients have the right to refuse treatment.

The NHS also commits to ensure those involved in patient care and treatment have access to health information to provide patient care safely and effectively. The NHS also pledges to anonymise the information collected during the course of treatment and use it to support research and improve care for others.

Informed choice: right to make choices about NHS care and to information to support these choices. What it means is that, increasingly, more information on the quality of services will be made available to the public.

Involvement in your healthcare and in the NHS: The right to be informed and have a say in one's own healthcare, but also to be involved in the planning of healthcare services. The constitution states that NHS organisations should work in partnership with service users, carers and the public, as well as with Local Authorities.

Complaint and redress: The right to complain, and have that complaint dealt with efficiently and be informed about the outcome. NHS organisations should commit to admitting to mistakes where they occur, explain what went wrong and put things right quickly and effectively.

3b. Patients and Public- responsibilities

- Please recognise that you can make a significant contribution to your own, and your family's, good health and wellbeing, and take personal responsibility for it.

- Please register with a GP practice.

- Please treat NHS staff and other patients with respect. You should recognise that abusive and violent behaviour could result in you being refused access to NHS services.

- Please provide accurate information about your health, condition and status.

- Please keep appointments, or cancel within reasonable time.

- Please follow the course of treatment which you have agreed, and talk to your clinician if you find this difficult.

- Please participate in important public health programmes such as vaccination.

- Please ensure that those closest to you are aware of your wishes about organ donation.

- Please give feedback – both positive and negative – about your experiences and the treatment and care you have received.

4a.Staff- rights and pledges

- have a good working environment with flexible working opportunities

- have a fair pay and contract framework;

- can be involved and represented in the workplace;

- have healthy and safe working conditions and an environment free from harassment, bullying or violence;

- are treated fairly, equally and free from discrimination;

- can in certain circumstances take a complaint about their employer to an Employment Tribunal; and

- can raise any concern with their employer, whether it is about safety, malpractice or other risk, in the public interest.

NHS also pledges:

The NHS commits:

- to provide a positive working environment

- to provide all staff with clear roles and responsibilities and rewarding jobs for teams and individuals that make a difference to patients, their families and carers and communities

- to provide all staff with personal development, access to appropriate education and training

- to provide support and opportunities for staff to maintain their health, wellbeing and safety

- to engage staff in decisions that affect them and the services they provide

- to have a process for staff to raise an internal grievance and

- to encourage and support all staff in raising concerns at the earliest reasonable opportunity about safety, malpractice or wrongdoing at work

4b. Staff- responsibilities

- You have a duty to accept professional accountability and maintain the standards of professional practice as set by your regulatory body

- You have a duty to take reasonable care of health and safety at work

- You have a duty to act in accordance with your contract of employment.

- You have a duty not to discriminate against patients or staff and to adhere to equal opportunities and equality and human rights legislation.

- You have a duty to protect the confidentiality of personal information that you hold.

- You have a duty to be honest and truthful in applying for a job and in carrying out that job.

Is the NHS constitution having any impact?

The NHS constitution for the first time, in one document, lays down the objectives of the NHS, the rights and responsibilities of the staff, patients and public and the guiding principles which govern the service. There is a statutory duty on NHS bodies, primary care services, and independent and third sector organisations providing NHS care in England to have regard to the NHS Constitution.

Yet, the future forum report in November 2012 said that despite the importance and potential of the NHS Constitution, its effect so far has been patchy, low key and inconsistent. It has failed to have the impact required to influence the quality of the service, the level of patient experience and give appropriate support to hard-working staff.

The main reasons cited for lack of impact were low levels of awareness about the constitution and lack of guidance as to what happens when the NHS falls short of people's expectations.

Thus, for the Constitution to succeed in its aims, it needs to become part of everyday life in the NHS for patients, the public and staff. Achieving this will require leadership, partnership and sustained commitment over months and years, to raise awareness of the Constitution and weave it into the way the NHS works at all levels. Publishing the Constitution is only the first step in the journey

Ref:

http://tinyurl.com/c9qmsac

NHS Litigation Authority (NHS LA)

The NHS LA was established in 1995 as a Special Health Authority. NHS LA is a not-for-profit part of the NHS. It provide indemnity cover for legal claims against the NHS, assist the NHS with risk management, share lessons from claims and provide other legal and professional services for its members.

Role of NHS LA:

- Managing claims against the NHS on behalf of its members. It settles justified claims fairly and quickly and defend unjustified claims robustly, helping to protect NHS resources
- Resolving disputes and claims fairly and cost effectively
- Helping the NHS to manage risk and reduce claims- NHS LA set standards for safe care and assess NHS providers against these. By ensuring members' contributions are fair and reflect risk NHS LA encourage them to provide safer care that reduces claims
- Professional advice-
    - Provide legal and professional advice to the NHS and to the Department of Health on a range of issues, including equal pay and age discrimination
    - Resolve contract disputes between health practitioners (including GPs, dentists, pharmacists and opticians) and their commissioners
    - National Clinical Assessment Service (NCAS) - helps resolve concerns about the professional practice of doctors, dentists and pharmacists in the UK.

Ref: www.nhsla.com

NHS Five Year Forward View

The NHS Five Year Forward View was published on 23rd Oct 2014. It sets out a vision for the future of the NHS. It was developed by the partner organisations that deliver and oversee health and care services including NHS England, Public Health England, Monitor, Health Education England, the Care Quality Commission and the NHS Trust Development Authority.

The purpose of the Five Year Forward View is to articulate why change is needed, what that change might look like and how we can achieve it. <u>This was the first time the NHS as a whole had set out its vision to government rather than vice versa</u>

Why change is needed?

The NHS had achieved considerable success in delivering efficiencies whilst maintaining services over recent years but this approach (e.g. pay restraint) could not be sustained indefinitely. Some of the fundamental challenges facing us:

- Increasing elderly population: we live longer, with complex health issues

- Modern advances in treatments and technologies- transforming our ability to predict, diagnose and treat disease

- Increasing budget pressures due to the global recession.

NHS England have previously predicted that if we continue with the current model of care and expected funding levels, we could have a funding gap of £30bn a year by 2020/21 which will continue to grow and grow quickly if action isn't taken.

The funding gap of £30bn supposes

- Uncontrolled rising demand
- No efficiency savings
- No additional funding

Therefore the three strands of a sustainable solution proposed in the 5 year forward plan are:

• Considerably greater emphasis on reducing demand through effective measures to prevent ill health (e.g. alcohol, obesity etc.) This will produce medium/longer terms benefits.

• Major changes in the models of care recognising the need for a path between a single centrally determined model and "letting a thousand flowers bloom" i.e. a limited menu of solutions to suit local needs.

•Additional funding from the government of £8bn a year.

Key themes in the NHS Five year forward view are:

Prevention

Radical upgrade in prevention and public health:

One in five adults still smoke. A third of us drink too much alcohol. A third of men and half of women don't get enough exercise. Just under two thirds of us are overweight or obese. The NHS will therefore now back hard-hitting national action on obesity, smoking, alcohol and other major health risks. We will help develop and support new workplace incentives to promote employee health and cut sickness-related unemployment. And we will advocate for stronger public health-related powers for local government and elected mayors.

New care models: Out of hospital care to be a larger part of what the NHS does

1. Patients will gain far greater control of their own care – including the option of shared budgets combining health and social care.

2. The NHS will take decisive steps to break down the barriers in how care is provided between family doctors and hospitals, between physical and mental health, between health and social care. The future will see far more care delivered locally but with some services in

specialist centres, organised to support people with multiple health conditions, not just single diseases.

3. England is too diverse for a 'one size fits all' care model to apply everywhere. But nor is the answer simply to let 'a thousand flowers bloom'. One new option will permit groups of GPs to combine with nurses, other community health services, hospital specialists and perhaps mental health and social care to create integrated out-of-hospital care - the Multispecialty Community Provider. A further new option will be the integrated hospital and primary care provider - Primary and Acute Care Systems - combining for the first time general practice and hospital services.

4. Across the NHS, urgent and emergency care services will be redesigned to integrate between A&E departments, GP out-of-hours services, urgent care centres, NHS 111, and ambulance services. Smaller hospitals will have new options to help them remain viable, including forming partnerships with other hospitals further afield, and partnering with specialist hospitals to provide more local services. Midwives will have new options to take charge of the maternity services they offer. The NHS will provide more support for frail older people living in care homes.

5. The foundation of NHS care will remain list-based primary care. Given the pressures they are under, we need a 'new deal' for GPs. Over the next five years, the NHS will invest more in primary care, while stabilising core funding for general practice nationally over the next two years. GP-led Clinical Commissioning Groups will have the option of more control over the wider NHS budget, enabling a shift in investment from acute to primary and

community services. The number of GPs in training needs to be increased as fast as possible, with new options to encourage retention.

The five year plan argues that there is nothing in their analysis that suggests that continuing with a comprehensive tax funded NHS is intrinsically undoable. The five year plan argues that delivering on the transformational changes set out in the plan and the resulting annual efficiencies could - if matched by staged funding increases as the economy allows – could close the £30 billion gap by 2020/21.

Ref: http://tinyurl.com/kcjenmc

NHS structure and Cash flows

Discuss the NHS management structure?

The Health and Social Care Act 2012 led to one of the biggest re organisation of NHS. The changes were aimed to:
- Improve the quality & choice of care for patients, & increase transparency for taxpayers;
- Give GPs & other clinicians the primary responsibility for commissioning health care;
- Create a coherent system of regulation for providers, to drive quality and efficiency;
- Limit Ministers' ability to micromanage, while ensuring they remain ultimately accountable.

Structure:

- The Secretary of State has overall responsibility for the work of the Department of Health (DH). DH provides strategic leadership for public health, the NHS and social care in England. Instead of directly managing providers or commissioners, Ministers will transparently set objectives for the NHS through a mandate to the NHS Commissioning Board. It will hold to account all of the national bodies, with powers to intervene in the event of significant failure, or in an emergency.

- NHS England is an independent body, at arm's length to the government. It provides national leadership for improving outcomes and driving up the quality of care by overseeing the operation of clinical commissioning groups (CCGs) and allocating resources to CCGs. It also commissions primary care and specialist services. NHS England also host clinical networks (to advise on single areas of care) and clinical senates (providing clinical advice on commissioning plans).

- Clinical commissioning groups (CCGs) are clinically led statutory NHS bodies responsible for the planning and commissioning of healthcare services for their local area. CCGs members include GPs and other clinicians such as nurses and consultants. They are responsible for about 60% of the NHS budget and commission most secondary care services such as: planned hospital care, rehabilitative care, urgent and emergency care (including out-of-hours), most community health services, mental health and learning disability services.

- Performance management: NHS providers will no longer be performance managed by Strategic Health Authorities. There will be a consistent dual system of regulation for all providers: the Care Quality Commission will regulate safety and quality while Monitor will act as economic regulator with powers to set prices, ensure competition works in patients' interests, and support service continuity.

- Patient and public involvement: New bodies known as local Health Watch will be established. They will help ensure that the views and feedback from patients and carers are an integral part of local commissioning across health and social care. Health Watch will be based in and funded by local authorities. A national body, Health Watch England, will be established to support local Health Watch. It will sit as a statutory committee of the CQC. Health Watch England will be tasked with representing people using health services at a national level and will have a role in advising CQC to review services where appropriate.

- Greater local accountability: New bodies called Health and Wellbeing Boards will be formed in all upper local tier local authorities to strengthen joint working between local government and the NHS. The boards will significantly increase local democratic legitimacy in the commissioning of health and care services, bringing together locally elected councillors, clinical commissioning groups, Local Health Watch and Directors of Adult Social Services, Children's Services and Public Health to jointly assess local needs and develop an integrated strategy to address them. Elected councillors will be involved in this process and will be held to account by the local electorate if they are ineffective. Local Health Watch will ensure patients and the public have a direct say in their health and wellbeing board and so in the strategic planning for meeting the health and care needs of their area.

- Monitor will temporarily also retain oversight of foundation trusts, while the NHS Trust Development Authority, a new body will be set up to oversee the transition of all NHS trusts to become foundation trusts.

- The National Institute for Health and Care Excellence will continue to provide independent advice and guidance to the NHS, and will extend its role to social care. The Information Centre will continue to act as the central, authoritative source of health and social care information

- Action to protect and promote the health of the population will be led nationally by a new public health service, Public Health England: an agency of the Department.

- Health Education England (HEE) was established as a Special Health Authority in June 2012 to promote high quality education and training as well as authorising and supporting Local Education and Training Boards (LETBs). It is expected that the education and training functions of SHAs and postgraduate deaneries will be undertaken by LETBs. These LETBs will provide a forum for workforce development to support research and innovation, co-ordinating workforce planning activity, and commissioning education and training locally.

NHS Structure (See the diagram)

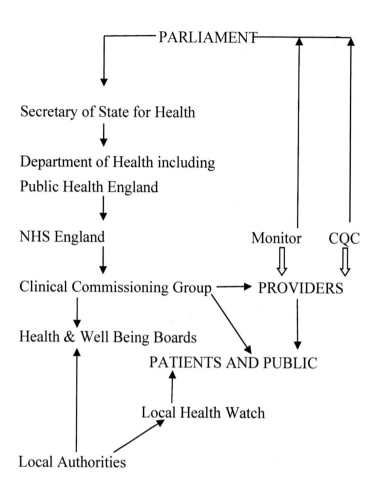

The NHS chief executives, Chief Medical Officer etc. work in the Department of Health and provide a link between the department of Health and the secretary of state for Health.

Discuss the cash flow in NHS.

The money for the NHS comes from the Treasury. Most of the money is raised through taxation. The Treasury holds a Spending Review every two to three years, through which the budget for NHS is agreed.

The Treasury allocates money to the Department of Health, which in turn allocates money to NHS England. The Department of Health retains a proportion of the budget for its running costs and the funding of bodies such as Public Health England. NHS England currently receives around £96 billion a year from the Department of Health (2012/13). Approximately £30 billion is retained by NHS England to pay for its running costs and the services it commissions directly: primary care (including GP services), specialised services, offender and military healthcare. The remainder is passed on to clinical commissioning groups (CCGs) to enable them to commission services for their populations. The NHS Mandate, issued annually from the government to NHS England, sets out what must be achieved in return for the taxpayer investment in the NHS.

How does NHS England decide how much each CCG gets? CCG budgets are allocated on a 'weighted capitation' basis. This means that budgets are set based on the size of the population, and adjusted for other factors: the age profile of the population; the health of the population; and the location of the population.

How is money paid to service providers? Historically, service providers were paid an annual lump sum to provide a service locally. These were known as 'block contracts', and were not linked to the number of patients seen, the work actually carried out, or the quality of care provided. In 2003/04 the government introduced 'Payment by Results' (PbR), an activity based system that reimburses providers for the work that they carry out, at an agreed national price. Currently, PbR represents almost 30% of NHS expenditure. Most of the remainder is covered by old-style block contracts and local variations on these. NHS England and local commissioners are working towards a payment system based on quality of care and health outcomes achieved.

Discuss the management structure of NHS hospitals.

Each NHS trust is headed by a board consisting of executive and non-executive directors, and is chaired by a non-executive director.

Non- Executive Directors:
- o These are lay people appointed form the local community served by the trust.
- o These non-executive directors bring their individual skills to the trust board.
- o The chairman of the trust board is a non-executive appointment.

Executive Directors:
- o These are experts led by a Chief Executive.
- o Each Executive has their functional responsibilities as well as being a corporate member of the Board.
- o The Executive Directors of the Trust Board are: Medical director, Finance director, Nursing director, Human resources director etc.

The clinical services are usually divided into 4 units (variously called Clinical business unit or Directorates). The Clinical Business units are Surgery, Medicine, Diagnostics and Women's and Child health. Each of them is managed by a Clinical director (usually a senior doctor) supported by the Services manager. The Divisions are responsible for the day-to-day management and delivery of services within their areas in line with Trust strategies, policies, and procedures.

National Institute for Health and Care Excellence

What is National Institute for Health and Care Excellence (NICE)? What do they do? How does NICE help you to look after your patients better? Tell us about a NICE guideline relating to your specialty and how it helps you.

NICE was set up as an independent organisation in 1999 to reduce variation in the availability and quality of NHS treatments and care - the so called 'postcode lottery'.

NICE's evidence-based guidance and other products help resolve uncertainty about which medicines, treatments, procedures and devices represent the best quality care and which offer the best value for money for the NHS. This helps better patient care.

As of 1 April 2013, NICE begins publishing social care guidance and quality standards to bridge the gap between health and social care services. **The National Institute for Health and Clinical Excellence has now become the National Institute for Health and Care Excellence.**

The Institute will still be known under the acronym NICE but the new name comes into effect to reflect the changes to its role as set out in the Health and Social Care Act (2012). As well as the Institute's increased remit, the Health and Social Care Act (2012) also puts the Institute on a stronger statutory footing, changing NICE from a Strategic Health Authority to a Non-Departmental Government Body.

Role of NICE

•**Guiding healthcare-** NICE develop evidence-based guidelines on the most effective ways to diagnose, treat and prevent disease and ill health.

•**Setting standards for high-quality care-** NICE is developing a library of 150 quality standards that will set out a vision of what high-quality care should look like on the NHS. These standards, broken down by a specific disease, condition or clinical area, will indicate when a treatment or procedure is considered highly effective, cost effective and safe, as well as being viewed as a positive experience by patients.

• **Develop quality standards for social care** (**from April 2013**) as part of their expanded remit (first two standards were published on supporting people with dementia to live well and on supporting the health and wellbeing of looked-after children and young people).

•**Recommendations on drugs**: NICE's technology appraisals programme is designed to ensure that people across England and Wales have equal access to new and existing medicines that are deemed clinically and cost effective, reducing the risk of a postcode lottery of care.

•**Public health:** Public health guidance is aimed at preventing ill health and encouraging people to live a healthy and active lifestyle. NICE have made recommendations on smoking, obesity and excessive alcohol consumption.

•**Diagnostics and medical technologies**: These are designed to evaluate medical diagnostics and open up access to new or innovative medical technologies and devices on the NHS.

•**Interventional procedures:** interventional procedures guidance evaluates the safety and efficacy of procedures used for diagnosis or treatment.

•**Support tools:** NICE provide support tools for the majority of its guidance to help put the recommendations into practice.

•**NHS Evidence:** NICE manage NHS Evidence, a service that provides access to authoritative clinical and non-clinical evidence and examples of best practice through a quick and easy online search engine.

•**Helping GPs provide high-quality care**: NICE oversees the development of indicators for the quality and outcomes framework (QOF), a voluntary incentive scheme that rewards GP practices across the UK for providing high-quality care to their patients.

•**Helping commissioners of health services**: Recent health reforms have outlined a role for NICE to work alongside the NHS Commissioning Board, and professional and patient groups, to develop a commissioning outcomes framework (COF). This will measure the health outcomes and quality of care achieved by clinical commissioning groups.

How does NICE work?

NICE develops guidance by a rigorous process that is centred on using the best available evidence and includes the views of experts, patients and carers, and industry.
NICE does not decide on the topics for its guidance and appraisals. Instead, topics are referred to NICE by the Department of Health.
Topics are selected on the basis of a number of factors, including the burden of disease, the impact on resources, and whether there is inappropriate variation in practice across the country.
The guidance is then created by independent and unbiased advisory committees.

How does NICE help me?

When NICE recommends a technology, the NHS must ensure it is available to those people it could help, normally within 3 months of the guidance being issued.

You are working as a consultant and you are recommending that one of your patients should be given a particular treatment based on the best evidence available. Hospital managers inform you that this treatment cannot be given as it is too costly. What do you do then? How would you inform the patient? What if the course of action you recommended was actually contained in a NICE guideline?

Every trust has a procedure for requesting treatment on a named patient basis for treatments which are not yet available in the trust. I will fill the relevant forms, approach the relevant committee & discuss it with my managers. The trust form basically asks you to justify the need for the treatment & evidence to support that the recommended treatment will be useful.
This normally resolves the issue. If not, I will inform the patient of the same as well as inform him how he can take the issue forwards.

Patient's perspective- if NICE recommended treatment not available.

•Patients can contact PALS for help and advice.

•Independent help and advice is also available from local Independent Complaints Advocacy Service. They have a statutory role to support patients and carers who wish to make a complaint about their NHS treatment or care.

•If the patient have been through the formal complaints process and received a final decision from your Trust, and are still unhappy, they can ask the Healthcare Ombudsman to investigate the complaint.

Ref: www.nice.org.uk

National Reporting and Learning System (NRLS)

NHS staff can report incidents to the NRLS through a specially designed electronic reporting form (known as the eForm) via NHS Net or the internet. They can also do so via their trust's local risk management system, from which incident data will be extracted and sent electronically to the NRLS. The aim of NRLS is not to investigate specific incidents, and this will remain the responsibility of the appropriate NHS bodies.

The NRLS is not a performance management, regulatory or investigative system, and is interested only in the 'how' and not the 'who'. As a result, the NRLS is confidential and anonymous and does not store anything that identifies either the reporter or patients/staff involved in a patient safety incident.

Through the NRLS, the aim is to develop an accurate picture of the extent of adverse incidents taking place in healthcare and have a baseline against which to measure improvements in patient safety. The aim is to use the NRLS data to seek to understand and tackle the "root causes" behind incidents and by sharing that learning help prevent the same incidents and errors occurring again.

Current status: NRLS has undoubtedly led to greater reporting of safety incidents, and is improving at feeding this information back to the NHS in a timely and useful way. Today, it is one of the most comprehensive such systems in the world.

NRLS is the world's most comprehensive database of patient safety information, to identify and tackle important patient safety issues at their root cause.

Given the hierarchical and blaming nature of healthcare, is it realistic to think that staff will be encouraged to point out their own mistakes, let alone those of a superior?

There are many barriers to creating a 'blame free' culture. However, many of the barriers are psychological; the fear of what someone else could do to the complainant. Blame free culture can only come about by helping NHS staff realise they don't need to feel threatened or feel guilty about reporting after they've done so. We've got to look after the staff and recognise the traumas that many staff experience from being involved in adverse events.

What is a patient safety incident?

A patient safety incident is defined as any unintended or unexpected incident which could have or did lead to harm for one or more patients receiving NHS-funded care.

National Patient Safety Alerting System (NPSAS)

This new patient safety system was launched in January 2014 for alerting the NHS to emerging patient safety risks. The new system allows for timely dissemination of relevant safety information to providers, as well as acting as an educational and implementation resource. It builds on the best elements of the former National Patient Safety Agency (NPSA) system.

Structure of NPSAS:

- A three-stage system, based on that used in other high risk industries.
- Used to disseminate patient safety information at different stages of development, to ensure newly identified risks can be quickly highlighted to providers.
- Allows rapid dissemination of urgent information, as well as encouraging information sharing between organisations and providing useful education and implementation resources for use by providers.

The three stage system of alerting:

Stage One Alert: Warning

This stage 'warns' organisations of emerging risk. It can be issued very quickly once a new risk has been identified to allow rapid dissemination of information.

Stage Two Alert: Resource

This alert may be issued some weeks or months after the stage one alert, and could consist of:

- sharing of relevant local information identified by providers following a stage one alert;
- sharing of examples of local good practice that mitigates the risk identified in the stage one alert;
- access to tools and resources that help providers implement solutions to the stage one alert; and

- access to learning resources that are relevant to all healthcare workers and can be used as evidence of continued professional development.

Stage Three Alert: Directive

When this stage of alert is issued, organisations will be required to confirm they have implemented specific solutions or actions to mitigate the risk.

What are the advantages of NPSAS?

- Gives organisations the opportunity to tackle emerging risk in their own way and to establish a sense of ownership.
- Through stage two alerts, organisations will be provided with potential solutions and resources to mitigate the risk.
- Encourages voluntary compliance for the early adopters, allowing providers to find solutions that best suit their individual organisations and minimises the requirement for directives.

All three stages of alert are likely to be used for issues representing a major risk. However, on occasions it may only be necessary to use part of the alert process. For example, issues of a widespread and well known nature may not require a Stage One: Warning, while those where a clear and specific solution exists may be addressed only with a Stage Three: Directive.

What are the consequences of failing to sign off stage one, two or three alerts by their deadline?

- Failure to comply is likely to be used by the CQC in their Intelligent Monitoring System and by commissioner's responsibilities for improving quality.
- Failure to comply with a Stage Three Alert: Directive within the deadline will be a cause for significant concern on the part of regulators, commissioners and most importantly, patients.

Ref: http://tinyurl.com/pcln8v4

National service framework (NSF)

How has the recent national service framework (NSF) affected your hospital's practice; how is its implementation being measured?

National service frameworks (NSFs) are long term strategies for improving specific areas of care. They set measurable goals within set time frames. NSFs set clear quality requirements for care. These are based on the best available evidence of what treatments and services work most effectively for patients.

National service frameworks (NSFs) cover some of the highest priority conditions. Currently there is NSF for cancer, coronary heart disease, diabetes, COPD, mental health, renal services, children, older people and long term conditions (neurological).

Thus, NSF has helped raise standards for the detection, treatment and management of cancer, heart diseases, diabetes, COPD etc.

NSFs:

- Set national standards and identify key interventions for a defined service or care group

- Put in place strategies to support implementation

- Establish ways to ensure progress within an agreed time scale

- Form one of a range of measures to raise quality and decrease variations in service in the NHS.

Each NSF is developed with the assistance of an external reference group (ERG) which brings together health professionals, service users and carers, health service managers, partner agencies, and other advocates. External reference groups adopt an inclusive process to engage the full range of views. The Department of Health supports the ERGs and manages the overall process.

Ref: http://tinyurl.com/6bzoe4z

NICE guidance on Nursing numbers

NICE issued guidance in July 2014 on safe staffing for nursing in adult inpatient wards in acute hospitals. This work was commissioned by the government in response to recommendations in the Francis inquiry into the mid-Staffordshire NHS trust.

NICE recommended that there is no single nursing staff-to-patient ratio that can be applied across the whole range of wards to safely meet patients' nursing needs. Each ward has to determine its nursing staff requirements to ensure safe patient care. This guideline therefore makes recommendations about the factors that should be systematically assessed at ward level to determine the nursing staff establishment.

But, it did not offer guidance on the minimum nurse staffing levels needed to deliver safe or high quality patient care. However, the guidance suggests that in many cases, patients' nursing needs, as determined by implementing the recommendations in this guideline, will require registered nurses to care for fewer than 8 patients.

Further from April 2014 all NHS Trusts are required to publish information about the nurse staffing levels. This was in response to Francis Report, which called for greater openness and transparency in the health service. All hospitals are now required to publish information about the number of nursing and midwifery staff working on each ward, together with the percentage of shifts meeting safe staffing guidelines.

Payment by results (PbR)

What do you understand of Payment by results (PbR) and national tariffs?

PbR is the payment system in England under which commissioners pay healthcare providers for each patient seen or treated, taking into account the complexity of the patient's healthcare needs. The two fundamental features of PbR are nationally determined currencies and tariffs. Currencies are the unit of healthcare for which a payment is made, and can take a number of forms covering different time periods from an outpatient attendance or a stay in hospital, to a year of care for a long term condition. Tariffs are the set prices paid for each currency.

PbR currently covers the majority of acute healthcare in hospitals, with national tariffs for admitted patient care, outpatient attendances, accident and emergency (A&E), and some outpatient procedures. For example, £119 for an outpatient attendance in obstetrics or £5,323 for a hip operation.

The currency for admitted patient care and A&E is the healthcare resource group (HRG). HRGs are clinically meaningful groups of diagnoses and interventions that consume similar levels of NHS resources. With some 26,000 codes to describe specific diagnoses and interventions, grouping these into HRGs allows tariffs to be set at a sensible and workable level. Under the latest version, HRG4, there are over 1,500 tariffs. Each HRG covers a spell of care, from admission to discharge. The currency for outpatient attendances is the attendance itself, divided into broad medical areas known as treatment function codes (TFCs).

Prior to PbR, hospitals were paid in locally negotiated block contracts. These allowed for considerable variations in prices for operations across the country, in line with their actual costs to hospitals. In addition hospitals were often paid even when they under-performed, failing to carry out number of operations required of them.

How does payment by results work?

When a patient is discharged, a clinical coder working in the hospital translates their care into codes using two classification systems, ICD-10 for diagnoses and OPCS-4 for interventions. When a patient attends an outpatient clinic, their TFC is similarly recorded. This information, together with other information about the patient such as age and length of stay, is sent from the hospital's computer system to a national database called the Secondary Uses Service (SUS). Reports from SUS allow commissioners to pay providers for the work they have done or to adjust any regular monthly payments for actual activity undertaken.

The tariff received by the provider is multiplied by a nationally determined market forces factor (MFF). This is unique to each provider and reflects the fact that it is more expensive to provide services in some parts of the country than in others. There may also be other adjustments to the tariff for long or short stays, for specialised services, or to support particular policy goals.

Why was PbR introduced?

Before PbR, commissioners tended to have block contracts with hospitals where the amount of money received by the hospital was fixed irrespective of the number of patients treated. PbR was introduced to:

- support patient choice by allowing the money to follow the patient to different types of provider;

- reward efficiency and quality by allowing providers to retain the difference if they could provide the required standard of care at a lower cost than the national price;

- reduce waiting times by paying providers for the volume of work done; and

- refocus discussions between commissioner and provider away from price and towards quality and innovation.

PbR was introduced to support healthcare policy and the strategic aims of the NHS. As these change and develop over time, so will PbR. The tariff is now seen increasingly as a vital means of supporting quality outcomes for patients and delivering additional efficiency in the NHS.

PbR began in a limited way, with national tariffs for 15 HRGs in 2003-04 and 48 HRGs in 2004-05. The first NHS foundation trust (FT) applicants moved to the full PbR system in 2005-06 and other NHS trusts in 2006-07. PbR now represents over 60% of acute hospital income.

Discuss the pros and cons of PbR.

Pros

- PbR aims to make the NHS more efficient and productive, undertaking more operations and treatments.

- Make the system more transparent about the work that hospitals actually do. Under the new system, hospitals will not get paid for unrecorded or badly recorded activity.

- Facilitate choice, by enabling funds to go he services chosen by patients.

Cons

- PbR rewards volume, not quality. Hospitals can make money if they bring costs down, or increase the amount of work they do. But cutting costs might be at the expense of better-quality equipment or staff numbers. However, PbR is being continually improved and there is a drive to link payment to quality by creating best practice tariffs. For example, the PbR team states that unnecessary follow up appointments take place in cataract pathways, and that the best practice tariff should therefore cover only the optimal amount.

- Another is that hospitals can start to cheat on coding. For instance, the NHS tariff pays two prices for different kinds of heart attack treatment: £1,775 for treatment of patients without medical complications, £3,676 for those with complications. The risk is that hospitals will falsify the code (or, worse still, give unnecessary treatment) in order to make more money.

- It is not easy to set tariffs for all health care activities. Procedures that have a clear treatment and roughly predictable length of stay - such as hip replacements – are relatively easy to cost, but it is much harder to set a fixed rate for treatments with fewer agreed definitions or end points. Some services and high cost drugs are currently excluded from PbR.

In summary, PbR has improved the fairness and transparency of the payment system. PbR has also possibly having a positive effect on activity and efficiency in elective care. However, PbR has not been in place long enough for any conclusions about its effectiveness to be drawn.

A recent Kings Fund report (Nov 2012) reviewed Payment by Results (PbR). The important findings from the report are as follows:

- While the introduction of PbR may have had some positive impacts within the NHS in England, the current system as applied is not fit for our current and future health and social care needs despite efforts to develop and refine it.

- PbR is not well designed to promote or support larger scale shifts in care from hospital to other settings due to incentives facing hospitals to maintain income and lack of flexibility to vary tariffs to reflect different costs of providing care in different settings.

- PbR is not well suited to promoting continuity and co-ordination of care. In its current form it does not provide payment relating to the costs of co-ordination itself and it does not provide a financial framework that supports or directly incentivises new ways of delivering care for people with long-term conditions.

- A single hospital episode such as an emergency admission may form just one part of an extended treatment cycle for some patients. Where the need for the episode is in part determined by the effectiveness of services in primary and community care, hospital treatment may not be required.

Although recent modifications to PbR have reduced the incentive to increase emergency admissions, they do not provide an incentive to reduce the underlying demand for this category of admission.

- For the NHS in England, the emphasis placed on giving greater priority to the prevention of illness, the treatment of people with long-term conditions, and the development of integrated care to address the needs of these people requires a radical rethink of the incentives needed.

- Attention should be given to the role of bundled payments that cover care for people with specific long-term conditions as well as those with co-morbidities, and the contribution of innovations such as year of care payments.

- More use could be made of capitated budgets to create incentives for providers to focus on prevention and on the provision of care in the most appropriate and cost effective settings.

- Different services require different payment systems. PbR is most appropriate to elective care and less suited to other services, where less rather than more activity is desirable, and where the nature of the service means that competition and choice is limited and the main requirement is to ensure there is the capacity to meet variable levels of demand.

- Different objectives means there will inevitably be trade-offs; the starkest of these is between cost and quality, and cost and maintenance of supply. High-quality standards and low prices, for example, could lead to limited supply.

- The impact of payment systems is still not well researched and data is limited.

The report recommends that the NHS Commissioning Board and Monitor should develop a payment strategy that is clear about the role and objectives of PbR. This should be part of a new framework that allows different payment systems for different types of service, and which also allows local flexibility on a clearly agreed basis.

Such a payment strategy should recognise how trade-offs between objectives will be handled, how robust data will be gathered for evaluation, and how that evaluation will be used to develop the system further.

Ref: http://www.kingsfund.org.uk/publications/payment-results-0

Under the terms of the Health and Social Care Act 2012, responsibility for currency and tariff design and price-setting for 2014-15 and beyond rests with the NHS Commissioning Board and Monitor.

PbR arrangements for 2013-14 build on the changes (and incorporating a lot of the critique above) made in recent years and continue to be guided by four key principles:

a) incentivising quality and better outcomes for patients

b) embedding efficiency and value for money within the tariff

c) promoting integration and patient responsiveness

d) expanding the scope of PbR.

Ref: http://tinyurl.com/pazhnah

Professional Standards Authority (PSA)

PSA is an independent statutory body covering all of the United Kingdom. It is answerable to Parliament. It was established by Parliament in 2003 to ensure consistency and good practice in healthcare regulation. It is responsible for overseeing UK's none health and care professional regulatory bodies including GMC.

PSA's mission is to promote the health & well-being of patients & the public in the regulation of health professionals by:

•helping regulatory bodies (GMC) become better regulators

•setting and driving up standards for professional regulation- by Sharing good practice & knowledge with the regulatory bodies, conducting research, & promoting the concept of right-touch regulation

•fostering greater harmonisation of regulatory practice & outcomes

•anticipating and influencing the future.

It does this by carrying out the following functions:

Checking how the regulators carry out their functions

Each year PSA carries out a performance review which looks at how each regulator (GMC) carries out its functions and their general performance against agreed standards. The reviews highlight good practice and identify issues that might benefit from a coordinated approach.

Referring cases to court

PSA looks at final stage decisions made by the regulators on professional's fitness to practise (FTP). It does this for all cases, except those where the health of the professional is under review. If PSA considers that a decision fails to protect the public interest, it has the power to investigate that decision and can refer it to the High Court, if needed.

Thus, PSA scrutinises and oversees the work of the nine regulatory bodies that set standards for training and conduct of health professionals.

Promoting good practice

PSA works with the regulators to improve quality and share good practice.

Advising health ministers

PSA can give advice to the Secretary of State and to the health ministers of Scotland, Wales and Northern Ireland about anything connected with a healthcare profession.

Why is PSA necessary?

It was created to help regulatory bodies become better regulators and to reassure the public that healthcare regulation is operating properly. A number of high-profile cases of poor practice and misconduct by healthcare professionals damaged public confidence in healthcare. It was set up by Parliament to oversee and co-ordinate the work of the regulators. It was given strong powers to ensure consistency and good practice in the public interest.

The Health and Social care Act 2012 gave PSA new powers by extending its remit to include social work in England and to accredit voluntary registers. PSA have been asked to set up a scheme to set standards for and accredit voluntary registers for people working within health and social care who are not required by law to be on one of the registers of the health and social work professional regulators in the UK.

Ref: www.professionalstandards.org.uk

QIPP- Quality, Innovation, Productivity and Prevention

QIPP is a large scale transformational programme for the NHS looking at how the NHS can deliver efficiency savings whilst maintaining or improving quality; it sets out the need to deliver improved services under tighter budget constraints, ever more important due to the current pressures on public sector budgets.

Only by driving up quality and productivity can the efficiency savings be realised to reinvest in meeting increasing demand and patient expectations. That is why the Quality, Innovation, Productivity and Prevention (QIPP) challenge has been and will continue to be of central importance.

NHS organisations at regional and local level all have QIPP plans in place to address the quality and productivity challenge, and supporting these are twelve national work streams designed to help NHS staff successfully deliver these changes.

The twelve national work streams are: Back office efficiency, Clinical support rationalisation, End of life care, Long term conditions, Medicines and procurement, Primary care commissioning, Productive care, QIPP procurement, Right care, Safe care and Urgent care.

There is also a lot of support and tools available to clinical teams and NHS organisations to help with the QIPP challenge. QIPP isn't an add-on, it encompasses the whole process of how the NHS can ensure sustainability in the way they fund care. It's about creating an environment in which change and improvement can flourish, and providing staff with the tools, techniques and support that will enable them to take ownership of improving quality of care.

Quality Accounts

Quality Accounts aim to enhance the public accountability of organisations that provide services to NHS patients and engage their leaders in quality improvement. Most providers of NHS services are required to annually produce and publish a Quality Account to give an account of the quality of those services and their priorities for improvement. Quality Accounts are currently not required in relation to primary care services and NHS Continuing Healthcare, or from organisations classed as 'small providers'.

The Quality Accounts look at:

- patient safety
- the effectiveness of treatments that patients receive
- patient feedback about the care provided

The Quality Accounts require the providers to answer series of questions-this includes information on how the healthcare provider goes about measuring how well it is doing, continuously improving the services it provides and how it responds to checks made by regulators such as the Care Quality Commission (CQC).

Revalidation and Appraisals

What do you know about Revalidation?

GMC introduced the licence to practise in 2009. To practise medicine in the UK all doctors are required by law to hold both registration and a licence to practise. Licensing was the first step towards the introduction of revalidation. Licences are required to be renewed periodically by revalidation.

Revalidation is the process by which doctors will have to demonstrate to the GMC, normally every five years, that they are up to date and fit to practise and complying with the relevant professional standards.

The purpose of this approach to medical regulation is to give patients a regular assurance that licensed doctors are up to date and fit to practise.

Revalidation started on 3 December 2012.

How does revalidation work?

Revalidation a step to step guide:

- Licensed doctors are required to link to a Responsible Officer (new statutory post). The responsible Officer is usually a senior, licensed doctor like the medical director in the healthcare organisation where the doctor works.

- Licensed doctors maintain a portfolio of supporting information drawn from their practice which demonstrates how they are continuing to meet the principles and values set out in Good Medical Practice Framework for appraisal and revalidation.

The supporting information needed for appraisal will fall under four broad headings:

- **General information** - providing context about what you do in all aspects of your work.

- **Keeping up to date** - maintaining and enhancing the quality of your professional work.

- **Review of your practice** - evaluating the quality of your professional work.

- **Feedback on your practice** - how others perceive the quality of your professional work.

- Licensed doctors undergo a process of annual appraisal based on their portfolio of supporting information.

- The Responsible Officer makes a recommendation to the GMC about a doctor's fitness to practise, normally every five years. The recommendation is based on the outcome of a licensed doctor's annual appraisals over the course of five years, combined with information drawn from the clinical governance system of the organisation in which the licensed doctor works.

- The GMC's decision to revalidate a licensed doctor is informed by the Responsible Officer's recommendation.

What are the key pieces of evidence a doctor needs to apply for revalidation?

Key pieces of evidence a doctor needs to apply for revalidation include:

- Evidence of continuous professional development

- Evidence of quality improvement activity

- Significant Events

- Colleague Feedback (multisource feedback)

- Patient feedback (where relevant)

- Review of complaints and compliments

Do you feel Revalidation is designed to resolve the issues they are meant to address?

The purpose of revalidation is to provide greater assurance to patients and the public, employers and other healthcare professionals that licensed doctors are up to date and fit to practise.

Revalidation aims to:

- Provide a focus for doctors' efforts to maintain and improve their practice

- Encourage the organisations in which doctors work to support their doctors to improve their practice and, where necessary, to identify and respond appropriately to emerging concerns about doctors at an early stage

- Encourage patients to provide feedback about the medical care they have received from a doctor, to be considered in their annual appraisals

In these ways, revalidation is contributing to the ongoing improvement in the quality of medical care delivered to patients throughout the UK. The principles of revalidation are sound. The process is likely to be effective as long as the trusts and doctors engage in the process.

Some people think that revalidation/appraisals are a waste of time and just a paperwork/box-ticking exercise. Do you feel it is a useful process?

The revalidation process requires the hospital and the consultant to reflect on their practice and the organisation and to identify possible improvements. As such it is a desirable process. However, it requires commitment from both the trust and the consultant. The trust need to provide training to its appraisers and appraisees and provide adequate time for it in the job plan. It also needs to support the consultant in achieving the objectives set in the PDP. Obviously, the consultant needs to be committed to the process for it to be effective.

What can you tell me about appraisals?

Appraisal is a formal process aimed to give doctors regular feedback on past performance, to chart their continuing progress and to identify education and development needs. It is part of a doctor's career development.

The key principles of professionalism set out in Good Medical Practice will be used to create a framework for annual appraisals. The evidence for appraisal will be collected under 4 domains:

- **General information** - providing context about what you do in all aspects of your work

- **Keeping up to date** - maintaining and enhancing the quality of your professional work

- **Review of your practice** - evaluating the quality of your professional work

- **Feedback on your practice** - how others perceive the quality of your professional work

The doctor and appraiser will agree a written overview of the appraisal, which should include a summary of achievement in the previous year, objectives for the next year, key elements of a personal development plan, actions expected of the organisation, a standard summary of the appraisal and a joint declaration that the appraisal has been carried out properly.

Personal development plan (PDP) - This is an outcome of the appraisal process listing the key development objectives of the appraisee for the following year as agreed with the appraiser.

What is the difference between Assessment and Appraisal?

Appraisal is a formal process to provide feedback on doctors' performance, chart their continuing professional development, and identify their developmental needs.

Assessment is a formal process which examines performance. In other words, assessment is ticking boxes set by others, whereas appraisal is ticking boxes that you have helped to set yourself. Revalidation will include both appraisal and assessment.

Who is the main beneficiary in an Appraisal or revalidation?

- It increases public confidence in doctors by reassuring the public that doctors are up to date and fit to practice.

- It leads to the personal and professional development of the individual and the NHS benefit as a whole.

What about appraisal of doctors in training?

Specialist training and progress through the grade are noted in the ARCP and are subject to assessment and development review.

Ref: http://www.gmc-uk.org/doctors/revalidation.asp

Seven Day Services

There is a growing movement towards more NHS services being available seven days a week.

Professor Sir Bruce Keogh, NHS England's National Medical Director is leading the drive for seven day services.

Key Questions:

Why move to 7 day services?

1. **Safer:** There are well rehearsed arguments that patient outcomes including mortality is worse at weekends (though data/analyses are complex).

2. **Convenience:** people would find it easier to access services outside the normal working week.

3. **Cost:** patient spending longer in hospital because of difficulties with weekend discharges. Others are admitted because of lack of rapid access to outpatient services at weekends.

What services are needed seven days a week?

It is important to differentiate between 24/7 and 7 day (but not out of hours) services. We also need to consider both primary and secondary care when talking about 7 day working.

Many services are already delivered 24/7 like A&E, hospital inpatients services (nursing and medical), some diagnostic services, out of hours GP services etc. while other health services like pharmacies are available 7 days a week (often for long hours though not 24/7)

Discuss the potential impact on outcomes of 7 day working?

Seven day working would lead to:

a. **Easier access to GP services.**

- May lead to patients (particularly working age) presenting early leading to better outcomes and reduced complications

- Reduction of A&E attendances/hospital attendances

- Greater convenience, improving patient experience

b. Easier access to hospital outpatients/day case services

- Rapid access clinics (TIA, chest pain, etc)
- Diagnostic services (imaging, endoscopy)
- Therapy (radiotherapy, day case surgery)

These would provide benefits in terms of convenience, patient experience, reduced admissions and possible impact on mortality and disability/quality of life. Further, running these services gives a critical mass of staff within the hospital to provide advice on other patients.

c. More senior staff leading care for inpatients would lead to:

- Better quality of decision making
- Better delivery of interventions
- Earlier discharge
- Fewer errors- reduced deaths and disability

What are the potential barriers to seven day working?

- Costs of additional staff
- Rota management
- Reluctance of staff to change work patterns
- Specialism vs. generalism
- Need to engage with social services
- Issues around bank holidays

How can progress be made?

- Leadership (winning hearts and minds)
- Planning and modelling
  - What will it mean for patients and carers?

- What will it mean for staff?
- What are the costs and benefits?
- Patient/public engagement
- Documentation/publication of existing good practice
- Testing and evaluation
- Commissioning
- Incentives
- Monitoring

Ref: http://tinyurl.com/qdj6jcd

Medical education and training - a new tariff system

A new system for funding medical education and training in England came into effect on 1 April 2014.

Why was a new system needed?

In the older system, payments to service providers (i.e. hospitals and GP surgeries) for the provision of clinical placements for students and trainees were based on local (and often historical) agreements, resulting in inequities in payment levels to providers within and between regions. Thus some providers were receiving more than the cost of placements with some receiving less.

The variation in levels of funding paid for medical education created an inequity between providers, with those receiving higher sums receiving an unfair advantage over those receiving lower sums. This resulted in cross-subsidisation of service from education and training money. There was little evidence that those receiving larger sums of money were providing placements of a higher quality.

Further, the funding arrangements for postgraduate medical education were not changed when the structure of the training programmes changed in 2007. The funding reflected assumptions that dated back over a decade on the amount of service a trainee provided at each stage of the old training structure. The funding arrangements needed updating to reflect the current training structures and the service contribution provided by trainee doctors.

How does the new tariff work?

The new funding arrangements pay the same price to all providers who are able to provide the same placements.

Local Education and Training Boards (LETBs- formerly called Deanery's) provide a lump sum payment to Local Education Providers (LEPs i.e. Hospitals) to cover the direct costs of delivering education and training for each placement for one year.

There are two components to the tariff:

Salary support - For 2014-15, salary support is 50% of the basic salary costs for the post across all grades. The remainder of the salary costs have to be met by the LEP.

Placement fee - The placement fee funds all the "direct costs" involved in delivering the education and training needs for an individual trainee. The fee for 2014-15 is £12,400 (multiplied by the Market Forces Factor).

Whistle blowing

The NHS constitution was updated in March 2012 to enshrine whistle blowing in law.

The updated Constitution includes:

• An expectation that staff should raise concerns at the earliest opportunity

• A pledge that NHS organisations should support staff when raising concerns by ensuring their concerns are fully investigated and that there is someone independent, outside of their team, to speak to

• Clarity around the existing legal right for staff to raise concerns about safety, malpractice or other wrong doing without suffering any detriment.

The above change is designed to make it easier for staff to raise concerns about poor patient care and hence improve patient safety.

The GMC too issued guidance on 'Raising and acting on concerns about patient safety' on 12th March 2012.

The publication of Raising and acting on concerns about patient safety is especially timely given the recent findings of the Mid Staffordshire enquiry.

Raising and acting on concerns about patient safety sets out that that all doctors have a duty to act when they believe patients' safety is at risk, or that patients' care or dignity is being compromised.

The guidance is separated into two parts.

Part 1: Raising a concern gives advice on raising a concern that patients might be at risk of serious harm, and on the help and support available to doctors.

Part 2: Acting on a concern explains doctors' responsibilities when colleagues or others raise concerns with them and how those concerns should be handled.

Part 1: Raising a concern

All doctors have a duty to raise concerns.

Overcoming obstacles to reporting:

You may be reluctant to report a concern for a number of reasons. For example, because you fear that nothing will be done or that raising your concern may cause problems for colleagues; have a negative effect on working relationships; have a negative effect on your career; or result in a complaint about you. If you are hesitating about reporting a concern for these reasons, you should bear the following in mind.

- You have a duty to put patients' interests first and act to protect them, which overrides personal and professional loyalties.
- The law provides legal protection against victimisation or dismissal for individuals who reveal information to raise genuine concerns and expose malpractice in the workplace.
- You do not need to wait for proof – you will be able to justify raising a concern if you do so honestly, on the basis of reasonable belief and through appropriate channels, even if you are mistaken.

Steps to raise a concern

- Local incident reporting systems
- Report to the regulatory body (GMC) in the following circumstances: a. If you cannot raise the issue with the responsible person or body locally because you believe them to be part of the problem. b. If you have raised your concern through local channels but are not satisfied that the responsible person or body has taken adequate action. c. If there is an immediate serious risk to patients, and a regulator or other external body has responsibility to act or intervene.

- Making a concern public: You can consider making your concerns public if you: a. have done all you can to deal with any concern by raising it within the organisation in which you work or which you have a contract with, or with the appropriate external body, and b. have good reason to believe that patients are still at risk of harm, and c. do not breach patient confidentiality.

Wherever possible, you should first raise your concern with your manager or an appropriate officer of the organisation you have a contract with or which employs you – such as the consultant in charge of the team, the clinical or medical director or a practice partner. If your concern is about a partner, it may be appropriate to raise it outside the practice – for example, with the medical director or clinical governance lead responsible for your organisation. If you are a doctor in training, it may be appropriate to raise your concerns with a named person in the LETB– for example, the postgraduate dean or director of postgraduate general practice education.

Part 2: Acting on concerns

All doctors have a responsibility to encourage and support a culture in which staff can raise concerns openly and safely. Concerns about patient safety can come from a number of sources, such as patients' complaints, colleagues' concerns, critical incident reports and clinical audit. Concerns may be about inadequate premises, equipment, other resources, policies or systems, or the conduct, health or performance of staff or multidisciplinary teams. If you receive this information, you have a responsibility to act on it promptly and professionally. You can do this by putting the matter right (if that is possible), investigating and dealing with the concern locally, or referring serious or repeated incidents or complaints to senior management or the relevant regulatory authority.

Who can help if you are not sure what to do? You can contact your medical defence body, royal college or BMA

Ref: http://tinyurl.com/ouobta6

Acute and emergency care: prescribing the remedy

Urgent and emergency care services face profound pressures that are most obviously experienced by patients and clinicians working in emergency departments and acute admission wards.

The Royal College of Physicians, the College of Emergency Medicine and the Royal Colleges of Surgeons and of Paediatrics and Child Health have produced a joint report, Acute and emergency care: prescribing the remedy, which provides 13 comprehensive local and national recommendations to address these challenges and to build safer, more effective and efficient urgent and emergency care services for all patients.

The recommendations are:

- Every emergency department should have a co-located primary care out-of-hours facility. It is unreasonable to expect patients to determine whether their symptoms reflect serious illness or more minor conditions. Co-location enables patients to be streamed following a triage assessment.

- Best practice that directs patients to the right care, first time, should be promoted across the NHS so as to minimise repetition of assessment, delays to care and unnecessary duplication of effort. Examples of best practice include: stroke patients being transferred directly to stroke Units, medical patients who have been assessed by a GP being taken directly to the medical admissions unit, patients with post-operative complications being returned to surgical care, GP-to-consultant advice lines, easy access to urgent clinics etc.

- All trainee doctors on acute specialty programmes should rotate though the emergency department.

- Senior decision-makers at the front door of the hospital, and in surgical, medical or paediatric assessment units, should be normal practice, not the exception. It should include acute physicians, acute paediatricians, GPs, emergency care physicians, geriatricians and psychiatrists.

- Emergency departments should have the appropriate skill mix and workforce to deliver safe, effective and efficient care. Where an emergency department does not have onsite back-up from particular specialties, there should be robust networks of care and emergency referral pathways.

- At times of peak activity, the system must have the capacity to deploy or make use of extra senior staff.

- Community and social care must be coordinated effectively and delivered 7 days a week to support urgent and emergency care services. The aim should be to facilitate the safe discharge and timely transfer of care of patients from the hospital to their own home or usual place of residence.

- Community teams should be physically co-located with the emergency department to bridge the gap between the hospital and primary and social care, and to support vulnerable patients. Co-located teams should include primary care practitioners, social workers and mental health professionals.

- The delivery of a seven-day service in the NHS must ensure that emergency medicine services are delivered 24/7, with senior decision makers and full diagnostic support available 24 hours a day, including appropriate access to specialist services. This will require additional resources.

- The funding and targets systems for emergency department attendances and acute admissions are unfit for purpose and require urgent change.

- Delivering 24/7 services requires new contractual arrangements that enable an equitable work–life balance.

- It is essential that each emergency department and acute admissions unit has an IT infrastructure that effectively integrates clinical and safeguarding information across all parts of the urgent and emergency care system.

- If configured properly with significant clinical involvement and advice, NHS 111, NHS 24, NHS Direct and equivalent telephone advice services can help to reduce the pressures on the urgent and emergency care system.

Ref: http://tinyurl.com/lkag58m

Mid-Staffordshire report (Francis report)

The Mid Staffordshire NHS Foundation Trust Public Inquiry was established in 2010 by the Secretary of State for Health to examine the commissioning, supervisory and regulatory organisations in relation to their monitoring role at Mid-Staffordshire NHS Foundation Trust between January 2005 and March 2009. The Inquiry was chaired by Robert Francis QC and considered why the serious problems at the Trust were not identified and acted on sooner and draw lessons to be learnt for the future of patient care.

The final report of the Inquiry was published on 6 February 2013. The report provides detailed and systematic analysis of what contributed to the failings in care at the trust. It recognises that what happened in Mid Staffs was a system failure, as well as a failure of the organisation itself. Rather than proposing a significant reorganisation of the system, the report concludes that a fundamental change in culture is required to prevent this system failure from happening again, and that many of the changes can be implemented within the current system. It stresses the importance of avoiding a blame culture, and proposes that the NHS – collectively and individually –adopt a learning culture aligned first and foremost with the needs and care of patients.

The report identifies that the failures of the Trust was primarily caused by a serious failure on the part of the Trust Board. It did not listen sufficiently to its patients and staff or ensure the correction of deficiencies was brought to the Trust's attention. Above all, it failed to tackle an insidious negative culture involving a tolerance of poor standards and a disengagement from managerial and leadership responsibilities. This failure was in part the consequence of allowing a focus on reaching national access targets, achieving financial balance and seeking foundation trust status to be at the cost of delivering acceptable standards of care.
The report says 'The story would be bad enough if it ended there, but it did not. The NHS system includes many checks and

balances which should have prevented serious systemic failure of this sort. There were and are a plethora of agencies, scrutiny groups, commissioners, regulators and professional bodies, all of whom might have been expected by patients and the public to detect and do something effective to remedy non-compliance with acceptable standards of care. For years that did not occur, and even after the start of the Healthcare Commission investigation, conducted because of the realisation that there was serious cause for concern, patients were, in my view, left at risk with inadequate intervention until after the completion of that investigation a year later. <u>In short, a system which ought to have picked up and dealt with a deficiency of this scale failed in its primary duty to protect patients and maintain confidence in the healthcare system'</u>.

The report has identified numerous warning signs which cumulatively, or in some cases singly, could and should have alerted the system to the problems developing at the Trust. That they did not has a number of causes, among them:

- A culture focused on doing the system's business – not that of the patients;
- An institutional culture which ascribed more weight to positive information about the service than to information capable of implying cause for concern;
- Standards and methods of measuring compliance which did not focus on the effect of a service on patients;
- Too great a degree of tolerance of poor standards and of risk to patients;
- A failure of communication between the many agencies to share their knowledge of concerns;
- Assumptions that monitoring, performance management or intervention was the responsibility of someone else;
- A failure to tackle challenges to the building up of a positive culture, in nursing in particular but also within the medical profession;

- A failure to appreciate until recently the risk of disruptive loss of corporate memory and focus resulting from repeated, multi-level reorganisation.

The **Francis report makes 290 recommendations** along the following themes:

1. Foster a common culture shared by all in the service of putting the patient first;
2. Develop fundamental standards and measures of compliance:
   - Develop a set of fundamental standards, easily understood and accepted by patients, the public and healthcare staff
   - Provide professionally endorsed and evidence based means of compliance with these fundamental standards which can be understood and adopted by the staff who have to provide the service;
   - Ensure that the relentless focus of the healthcare regulator is on policing compliance with these standards. Non-compliance with these standards should not be tolerated and any organisation not able to consistently comply should be prevented from continuing a service which exposes a patient to risk
   - To cause death or serious harm to a patient by non-compliance without reasonable excuse of the fundamental standards <u>should be a criminal offence</u>
   - These fundamental standards should be policed by the Care quality commission (CQC)
   - The merger of the regulation of care into one body – with Monitor responsibilities being absorbed by the CQC over time

3. Ensure openness, transparency and candour throughout the system underpinned by statute. Without this a common culture of being open and honest with patients and regulators will not spread.
   - The " duty of candour" - a statutory duty to be truthful to patients where harm has or may have been caused

- Staff to be obliged by statute to make their employers aware of incidents in which harm has been or may have been caused to a patient
- Trusts have to be open and honest in their quality accounts describing their faults as well as their successes
- The deliberate obstruction of the performance of these duties and the deliberate deception of patients and the public should be a criminal offence
- It should be a criminal offence for the directors of Trusts to give deliberately misleading information to the public and the regulators
- The CQC should be responsible for policing these obligations

4. Enhance the recruitment, education, training and support of all the key contributors to the provision of healthcare, but in particular those in nursing and leadership positions, to integrate the essential shared values of the common culture into everything they do;

- Improved support for compassionate, caring and committed nursing
  - Entrants to the nursing profession should be assessed for their aptitude to deliver and lead proper care, and their ability to commit themselves to the welfare of patients
  - Training standards need to be created to ensure that qualified nurses are competent to deliver compassionate care to a consistent standard
  - Nurses need a stronger voice, including representation in organisational leadership and the encouragement of nursing leadership at ward level
  - Healthcare workers should be regulated by a registration scheme, preventing those who should not be entrusted with the care of patients from being employed to do so.
- Stronger healthcare leadership
  - The establishment of an NHS leadership college, offering all potential and current leaders the

chance to share in a common form of training to exemplify and implement a common culture, code of ethics and conduct
- o It should be possible to disqualify those guilty of serious breaches of the code of conduct or otherwise found unfit from eligibility for leadership posts
- o A registration scheme and a requirement need to be established that only fit and proper persons are eligible to be directors of NHS organisations.

So in summary, the Francis report recommends making all those who provide care for patients – individuals and organisations – properly accountable for what they do and to ensure that the public is protected from those not fit to provide such a service. It also recommends development and sharing of ever-improving means of measuring and understanding the performance of individual professionals, teams, units and provider organisations for the patients, the public, and all other stakeholders in the system.
Ref: http://www.midstaffspublicinquiry.com/report

Government Response

The government published a **full response on 19th Nov 2013** to the 290 recommendations made by Robert Francis. In total, the Government accepted 281 out of 290 recommendations from the inquiry's report.

The key actions from the government response include:

1. Transparent, monthly reporting of ward-by-ward staffing levels and other safety measures. But the Government's response stopped short of introducing a minimum staff-patient ratio on wards or enshrining this in law because staff requirements were a "different number for different wards".
2. Quarterly reporting of complaints data and lessons learned by trusts along with better reporting of safety incidents

3. A statutory duty of candour on providers, and professional duty of candour on individuals, through changes to professional codes.

There will be no statutory duty of candour on individual NHS staff to tell patients or their families if incidents have led to serious harm or death. Instead, the Government will impose such a duty on organisations as a whole and will strengthen duty of candour on individuals using organisations such as the GMC.

4. Trusts to be liable if they have not been open with a patient. Hospitals will lose their insurance cover for that case if they are not. Trusts will be liable if they have not been open with a patient – the NHS Litigations Authority will continue to make full payments on successful claims, but will have discretion to make the trust partly liable

5. A dedicated hospital safety website to be developed for the public. This website will draw together up to date information on all the factors, for which robust data is available, that impact on the safety of care

6. A new national patient safety programme across England to spread best practice and build safety skills across the country and 5,000 patient safety fellows will be trained and appointed in 5 years

7. A new criminal offence for wilful neglect, with a government intention to legislate so that those responsible for the worst failures in care are held accountable

8. A new fit and proper person test, to act as a barring scheme for senior managers

The public have the right to expect that people in leading positions in NHS organisations are fit and proper persons; and that where it is demonstrated that a person is not fit and proper (i.e. good character including past employment history, has the qualifications, skills and experience necessary for the work or office as well as the more traditional consideration of criminal and financial matters) they should not be able to occupy such a position. Monitor and the Care Quality Commission are committed to ensuring that, taken together, their processes for registration and licensing reflect these principles. The Care Quality Commission's inspection regime will include a focus on whether or not an organisation is 'well-led'.

9. Every hospital patient to have the names of a responsible consultant and nurse above their bed

10. A named accountable clinician for out-of-hospital care for all vulnerable older people.

11. More time to care as all arms' length bodies and the Department of Health have signed a protocol in order to minimise bureaucratic burdens on trusts

12. A new care certificate to ensure that healthcare assistants and social care support workers have the right fundamental training and skills

13. A new fast-track leadership programme to recruit clinicians and external talent to the top jobs in the NHS in England.

Ref: http://francisresponse.dh.gov.uk/

Future hospital: caring for medical patients

Our hospitals are struggling to cope with the challenge of an ageing population and increasing hospital admissions. All too often our most vulnerable patients – those who are old, who are frail or who have dementia – are failed by a system ill-equipped and seemingly unwilling to meet their needs.

Issues with the current system:

- a health system ill-equipped to cope with the needs of an aging population with increasingly complex clinical, care and support needs
- hospitals struggling to cope with an increase in clinical demand
- a systematic failure to deliver coordinated, patient-centred care, with patients forced to move between beds, teams and care settings with little communication or information sharing
- services that struggle to deliver high-quality services across 7 days, particularly at weekends
- a looming crisis in the medical workforce, with consultants and medical registrars under increasing pressure, and difficulties recruiting to posts and training schemes that involve general medicine.

The Future Hospital Commission report '**Future hospital: caring for medical patient's**' 2013 by the Royal College of Physicians (RCP) is an attempt to deal with the problems above. It sets out the Commission's vision for hospital services structured around the needs of patients.

Creating the future hospital

1. A new principle of care

The Future Hospital Commission sets out a radical new model of care designed to encourage collective responsibility for the care of patients across professions and healthcare teams. Care should come to patients and be coordinated around their medical and support needs. However, it is not unusual for patients – particularly older people – to move beds several times during a single hospital stay. This results in poor care, poor patient experience and increases length of stay. In the future hospital, moves between beds and wards will be minimised and only happen when this is necessary for clinical care.

Delivery of specialist medical care – such as cardiology and neurology services – will not be limited to patients in specialist wards or to those who present at hospital. Specialist medical teams will work cross the whole hospital and out into the community across 7 days.

2. A new model of care

To coordinate care for patients, the Future Hospital Commission recommends that each hospital establish the following new structures.

a) Medical Division - it will be responsible for all medical services across the hospital – from the emergency department and acute and intensive care beds, through to general and specialist wards. Medical teams across the Medical Division will work together to meet the needs of patients, including patients with complex conditions and multiple comorbidities. The Medical Division will work closely with partners in primary, community and social care services to deliver specialist medical services across the health economy.

The Medical Division will be led by the chief of medicine, a senior doctor responsible for making sure working practices facilitate collaborative, patient-centred working and that team's work together towards common goals and in the best interests of patients.

b) Acute Care Hub - it will bring together the clinical areas of the Medical Division that focus on the initial assessment and stabilisation of acutely ill medical patients. These include the acute medical unit, the ambulatory care centre, short-stay beds, intensive care unit and, depending on local circumstances, the emergency department. The Acute Care Hub will focus on patients likely to stay in hospital for less than 48 hours, and patients in need of enhanced, high dependency or intensive care.

An acute care coordinator will provide operational oversight to the Acute Care Hub, supported by a Clinical Coordination Centre.

c) Clinical Coordination Centre - it will be the operational command centre for the hospital site and Medical Division, including medical teams working into the community. It will provide healthcare staff with the information they need to care for patients effectively. It will hold detailed, real-time information on patients' care needs and clinical status, and coordinate staff and services so that they can be met. In the longer-term, this would evolve to include information from primary and community care, mental health and social care. This information would be held in a single electronic patient record, developed to common standards.

3. Seven-day care, delivered where patients need it

We must bring the advances in medical care to all patients, whatever their additional needs and wherever they are in hospital or the community. This means specialist medical teams will work – not only in specialist wards – but across the hospital. Care for patients with multiple conditions will be coordinated by a single named consultant, with input from a range of specialist teams when patients' clinical needs require it. The remit and capacity of medical teams will extend to adult inpatients with medical problems across the hospital, including those on 'non-medical' wards (e.g. surgical patients).

Once admitted to hospital, patients will not move beds unless their clinical needs demand it. Patients should receive a single initial assessment and ongoing care by a single team. In order to achieve this, care will be organised so that patients are reviewed by a senior doctor as soon as possible after arriving at hospital. Specialist medical teams will work together with emergency and acute medicine consultants to diagnose patients swiftly, allow them to leave hospital if they do not need to be admitted, and plan the most appropriate care pathway if they do. Patients whose needs would best be met on a specialist ward will be identified swiftly so that they can be 'fast-tracked' – in some cases directly from the community.

When a patient is cared for by a new team or moved to a new setting, there will be rigorous arrangements for transferring their care (through 'handover'). This process will be prioritised by staff and supported by information captured in an electronic patient record that contains high-quality information about patients' clinical and care needs.

Specialist medical care will not be confined to inside the hospital walls. Medical teams will work closely with GPs and those working in social care to make sure that patients have swift access to specialist care when they need it, wherever they need it. Much specialised care will be delivered in or close to the patient's home. Physicians and specialist medical teams will expect to spend part of their time working in the community, with a particular focus on caring for patients with long-term conditions and preventing crises.
To support this way of working, the performance of specialist medical teams will be assessed according to how well they meet the needs of patients with specified conditions across the hospital and health economy, not just those located on specialist wards.

Acutely ill medical patients in hospital should have the same access to medical care on the weekend as on a week day. Services should be organised so that clinical staff and diagnostic and support services are readily available on a 7-day basis. The level of care available in hospitals must reflect a patient's severity of illness. In order to meet the increasingly complex needs of patients – including those who have dementia or are frail – there will be more beds with access to higher intensity care, including nursing numbers that match patient requirements.

There will be a consultant presence on wards over 7 days, with ward care prioritised in doctors' job plans. Where possible, patients will spend their time in hospital under the care of a single consultant-led team. Rotas for staff will be designed on a 7-day basis, and coordinated so that medical teams work together as a team from one day to the next.

Care for patients should focus on their recovery and enabling them to leave hospital as soon as their clinical needs allow. This will be planned from when the patient is admitted to hospital and reviewed throughout their hospital stay. Arrangements for patients leaving hospital will operate on a 7-day basis. Health and social care services in the community will be organised and integrated to enable patients to move out of hospital on the day they no longer require an acute hospital bed.

Patients can be empowered to prevent and recover from ill health through effective communication, shared decision-making and self-management.

Patients should only be admitted to hospital if their clinical needs require it. For many, admission to hospital is the most effective way to set them on the road to recovery. However, it can be disorientating and disruptive. In the future, hospitals will promote ways of working that allow emergency patients to leave hospital on the same day, with medical support provided outside hospital if they need it.

Doctors will assume clinical leadership for safety, clinical outcomes and patient experience. This includes responsibility to raise questions and take action when there are concerns about care standards, and collaborate with other teams and professions to make sure that patients receive effective care throughout the hospital and wider health and care system.

There will always be a named consultant responsible for the standard of care delivered to each patient. Patients will know who is responsible for their care and how they can be contacted. The consultant will be in charge of coordinating care for all patients on the ward, supported by a team. The consultant and ward manager will assume joint responsibility for ensuring that basic standards of care are delivered, and that patients are treated with

dignity and respect. Nurse leadership and the role of the ward
manager will be developed and promoted.

There will be mechanisms for measuring patients' experience
of care. This information will be used by hospitals, clinical teams
and clinicians to reflect on their practice and drive improvement.
A Citizenship Charter that puts the patient at the centre of
everything the hospital does should be developed with patients,
staff and managers. This should be based on the NHS
Constitution and embed in practice the principles of care set out
by the Future Hospital Commission.

4. Education, training and deployment of doctors
Medical education and training will develop doctors with the
knowledge and skills to manage the current and future
demographic of patients. We need a cadre of doctors with the
knowledge and expertise necessary to diagnose, manage and
coordinate continuing care for the increasing number of patients
with multiple and complex conditions. This includes the expertise
to manage older patients with frailty and dementia. Across the
overall physician workforce there will be the skills mix to deliver
appropriate:
- specialisation of care – access to sufficient specialty
  expertise to deliver diagnosis, treatment and care
  appropriate to the specific hospital setting
- intensity of care – access to sufficient expertise to
  manage, coordinate and deliver enhanced care to patients
  with critical illness
- co-ordination of care – access to sufficient expertise to
  coordinate care for patients with complex and multiple
  comorbidity.

In order to achieve the mix of skills that delivers for patients, a
greater proportion of doctors will be trained and deployed to
deliver expert (general) internal medicine care. The importance
of acute and (general) internal medicine must be emphasised
from undergraduate training onwards, participation in (general)
internal medicine training will be mandatory for those training in

all medical specialties, and a more structured training programme for (general) internal medicine will be developed.

The contribution of medical registrars will be valued and supported by increased participation in acute services and ward-level care across all medical trainees and consultants, and enhanced consultant presence across 7 days.

Ref: http://tinyurl.com/nsd9pdt

Keogh Urgent and Emergency Care Review

High quality care for all, now and for future generations:
Transforming urgent and emergency care services in England.

Urgent and Emergency Care Review by Sir Bruce Keogh, The
National Medical Director of NHS England- Phase 1 review
published in Nov 2013 (This review is being conducted in 2
phases).

Background/case for change:

There was a real concern that A&E departments, hospital services
supporting A&E and ambulance services were under intense,
growing and unsustainable pressure.
The reasons for the growing pressures our A&E departments are
well rehearsed as follows: Firstly, an ageing population with
increasingly complex needs is leading to ever rising numbers of
people needing urgent or emergency care. Secondly, many people
are struggling to navigate and access a confusing and inconsistent
array of urgent care services (urgent care centre, walk in centres,
minor injuries unit etc) provided outside of hospital, so they
default to A&E.

While both these things are true, they arguably underplay the fact
that A&E departments have become victims of their own success.
The A&E brand is trusted by the public and, despite increasing
pressure, continues to provide a very responsive service with an
average wait for treatment of only 50 minutes and the
overwhelming majority of patients being treated within 4 hours.
So, we should not be surprised that people choose to go to A&E.
But, the reality is that millions of patients every year seek or
receive help for their urgent care needs in hospital who could
have been helped much closer to home. The opportunities for
bringing about a shift from hospital to home are enormous. For
example, we know that 40% of patients attending A&E are
discharged requiring no treatment at all; there were over 1 million
avoidable emergency hospital admissions last year; and up to 50

per cent of 999 calls requiring an ambulance to be dispatched could be managed at the scene. To seize the opportunities these numbers present, we will need to greatly enhance urgent care services provided outside of hospital. This forms a key part of our proposals. **The comprehensive review was conducted into how to organise and provide urgent an emergency care services in England.**

Accident and Emergency departments
In the 1970s most A&Es and their hospitals could offer people the best treatment of the day for most conditions. Clinical practice has taken great strides forward in the last four decades, and this is no longer the case for e.g. not all A&E departments offer stroke thrombolysis or emergency coronary revascularilisation. Advancing science has directed the way we deliver services to achieve the best results, but it also exposes the illusion that all A&Es are equally able to deal with anything that comes through their doors. We now find ourselves in a place where, unwittingly, patients have gained false assurance that all A&E's are equally effective. This is simply not the case. Many A&E departments do not have consistent consultant presence overnight or at weekends. The support services available also vary considerably, with 1 in 7 lacking at least one "essential" on-site service, such as critical care, acute medicine, acute surgery or trauma and orthopaedics.

So, A&E departments up and down the country offer very different types and levels of service, yet they all carry the same name. We need to ensure that there is absolute clarity and transparency about what services different facilities offer and direct or convey patients to the service that can best treat their problem. Most importantly, we need to ensure that anywhere that displays a red and white sign is a place that will provide access to the very best care for the most seriously ill and injured patients, 24 hours a day and 7 days a week. A place that can resuscitate, make a diagnosis, start treatment and ensure rapid transfer to the right place if it can't offer the very best care.

Vision in the report:
Sir Bruce says: "Our vision is simple.
Firstly, for those people with urgent but non-life threatening needs we must provide highly responsive, effective and personalised services outside of hospital. These services should deliver care in or as close to people's homes as possible, minimising disruption and inconvenience for patients and their families.
Secondly, for those people with more serious or life threatening emergency needs we should ensure they are treated in centres with the very best expertise and facilities in order to reduce risk and maximise their chances of survival and a good recovery. If we can get the first part right then we will relieve pressure on our hospital based emergency services, which will allow us to focus on delivering the second part of this vision."

The report makes proposals in five key areas:

1. Providing better support for people to self-care – The NHS will provide better and more easily accessible information about self-treatment options so that people who prefer to can avoid the need to see a healthcare professional.
2. Helping people with urgent care needs to get the right advice in the right place, first time – The NHS will enhance the NHS 111 service so that it becomes the smart call to make, creating a 24 hour, personalised priority contact service. This enhanced service will have knowledge about people's medical problems, and allow them to speak directly to a nurse, doctor or other healthcare professional if that is the most appropriate way to provide the help and advice they need. It will also be able to directly book a call back from or an appointment with, a GP or at whichever urgent or emergency care facility can best deal with the problem.
3. Providing highly responsive urgent care services outside of hospital so people no longer choose to queue in A&E - This will mean: putting in place faster and consistent same-day, every-day access to general practitioners,

primary care and community services such as local mental health teams and community nurses to address urgent care needs; harnessing the skills, experience and accessibility of community pharmacists; developing our 999 ambulance service into a mobile urgent treatment service capable of treating more patients at scene so they don't need to be conveyed to hospital to initiate care.

4. Ensuring that those people with more serious or life threatening emergency needs receive treatment in centres with the right facilities and expertise in order to maximise chances of survival and a good recovery. Once it has enhanced urgent care services outside hospital, the NHS will introduce two types of hospital emergency department with the current working titles of Emergency Centres and Major Emergency Centres. In time, these will replace the inconsistent levels of service provided by A&E Departments. The presence of senior clinicians seven days a week will be important for ensuring the best decisions are taken, reassuring patients and families and making best use of NHS resources. Emergency Centres will be capable of assessing and initiating treatment for all patients and safely transferring them when necessary. Major Emergency Centres will be much larger units, capable of not just assessing and initiating treatment for all patients but providing a range of highly specialist services. The NHS envisages around 40-70 Major Emergency Centres across the country. It expects the overall number of Emergency Centres – including Major Emergency Centres – carrying the red and white sign to be broadly equal to the current number of A&E departments.

5. Connecting urgent and emergency care services so the overall system becomes more than just the sum of its parts. Building on the success of major trauma networks, the NHS will develop broader emergency care networks. These will dissolve traditional boundaries between hospital and community-based services and support the free flow of information and specialist expertise. They will ensure that no contact between a clinician and a

patient takes place in isolation – other specialist expertise will always be at hand. Major Emergency Centres will have a lead responsibility for the quality of care and operational performance of services across the network they support, including linked Emergency Centres.

Sir Bruce adds: "Let me be clear that there is no simple solution. This report sets out some principles. How they are developed locally will, and must, vary to suit local circumstances and wishes. We will need different approaches in metropolitan, rural or remote areas. We know people will want to see change as soon as possible, but we need to ensure that there are no risky, ill considered "big bangs", and that there is a managed transition." We anticipate that it will take 3-5 years to enact the major transformational change set out within this report.

Ref: http://tinyurl.com/prx9qtg

Phase 2 of the review: the delivery phase

The second, delivery phase of the review (i.e. 'what' needs to change into 'how' change can and will be delivered) aims to convert the work done so far into a framework that will guide and support commissioners, clinicians and providers in the transformation of urgent and emergency care services.

An Urgent and Emergency Care Delivery Group has been established to work out the practicalities for delivering the whole-system change.

The Delivery Group has identified eight key areas where work is needed, and subgroups have been established to carry out the necessary work in each of these areas. The eight areas are:

- Whole system planning and payment, commissioning and accountability

- Primary and community care access

- 111 (contact first/smart call)

- Data, information and care planning

- Community pharmacy

- Emergency departments and emergency care networks

- Ambulance treatment service

- Workforce

The Delivery group highlighted that it would take three to five years to enact the major transformational changes set out in the report, but expected to make significant progress over the next six months in five specific areas. A summary of progress was reported in Aug 2014 against each of these five specific areas:

- Working closely with local commissioners as they develop their five year strategic and two year operational plans: Planning guidance was issued to commissioners to stimulate thought as to how they should prepare the way for the establishment of the urgent and emergency care networks envisaged in our End of Phase 1 Report
- Identifying and initiating transformational demonstrator sites to trial new models of delivery for urgent and emergency care and 7 day services, supported by NHS Improving Quality (NHSIQ)
- Developing new payment mechanisms for urgent and emergency care services, in partnership with Monitor
- Completion of the new NHS 111 service specification so the revised service (which will go live during 2015/16) can meet the aspirations of this Review
- Working through the NHS Commissioning Assembly to develop and co-produce with CCGs the necessary commissioning guidance and specifications for new ways of delivering urgent and emergency care (with this process continuing over the remainder of 2014/15)

Ref: http://tinyurl.com/mqt4lpb

Shape of Training Review

Shape of Training Review: A new way of training doctors for a changing healthcare landscape

Professor David Greenaway, Vice-Chancellor of Nottingham University, was asked to review UK postgraduate medical education and training in March 2012. The review was jointly sponsored by the Academy of Medical Royal Colleges, the General Medical Council, Medical Education England, the Medical Schools Council, NHS Scotland, NHS Wales and the Northern Ireland Department of Health, Social Services and Public Safety. **The review was published in Oct 2013.**

The purpose of the review was to shape the way doctors are trained in the UK to make sure they are equipped to meet the changing needs of patients, society and health services.

Need for the review:

The needs of patients in the UK are changing fast. There is an increasingly ageing population with multiple co-morbidities. To ensure that doctors have the required skills aptitudes to meet our changing needs, it was felt that a rethink about current arrangements for postgraduate medical education and training was needed. In particular, that we need a better balance between doctors who are trained to provide care across a general specialty area and those prepared to deliver more specialised care.

The main recommendations were:

1. Patients and the public need more doctors who are capable of providing general care in broad specialties across a range of different settings.

The review describes an approach to training in the future that will develop more broadly trained specialists. The key milestones in this model are outlined below:

a. Full registration should happen at the point of graduation from medical school. Measures will need to be put in place to make sure graduates are fit to work as fully registered doctors.

b. Following graduation, doctors will undertake the two-year Foundation Programme.

c. After the Foundation Programme, doctors will enter broad based specialty training. Specialties or areas of practice will be grouped together. These groupings will be characterised by patient care themes (such as women's health, child health and mental health), and will be defined by the dynamic and interconnected relationships between the specialties. They will have common clinical objectives, set out in the specialty curricula.

d. Broad based specialty training, after Foundation Programme, will last between four and six years depending on specialty requirements.

e. During postgraduate training, doctors should be given opportunities to spend up to a year working in a related specialty or undertaking education, leadership or management work (similar to specialty fellowships). This year, which can be taken at any time during training, will allow them to gain wider experiences that will help them become more rounded professionals. It will be included in the timeframe of between four and six years.

f. The exit point of postgraduate training will be the Certificate of Specialty Training. It marks the point at which doctors are able to practise in their identified scope of practice, with no clinical supervision, while working in multi-professional teams.

g. Most doctors will work in the general area of their broad specialty, based on patient and workforce needs, throughout their careers. They will be expected to maintain and develop their skills in their specialty area.

h. All doctors must be able to manage acutely ill patients with multiple co-morbidities within their broad specialty areas

i. The placements for training should be longer and training should be based on apprenticeship based model.

2. We will continue to need doctors who are trained in more specialised areas to meet local patient and workforce needs.

3. Training should be limited to places that provide high quality training and supervision, and that are approved and quality assured by the GMC.

4. All doctors must be able to manage acutely ill patients with multiple co-morbidities within their broad specialty area and, continue to maintain their skills in the future.

<u>Other relevant issues covered in the review:</u>

Moving full registration to the point of graduation: Currently the support and management of F1 doctors is fragmented. Medical schools are responsible for considering their fitness to practise and making recommendations to the GMC about full registration. But F1 training can take place anywhere in the UK. F1 doctors have little or no supervisory relationship with their medical school, while postgraduate organisations face challenges in managing F1 doctors who have fitness to practise concerns due to complex governance arrangements. By moving full registration to the point of graduation, responsibility for F1 doctors will clearly be with postgraduate institutions.

However, moving the point of full registration will be complex. Patients and the service are likely to expect graduates who have full registration to meet the same competence level as the current threshold. This change will inevitably have a knock-on effect on undergraduate medical education, which will have to ensure graduates meet more advanced outcomes. Further, we should move towards limiting where doctor's train following graduation to places that provide high quality training and supervision, and that are approved and quality assured by the GMC. They must not work in places just because there is a service need for doctors in training.

Blurring the boundary between primary and secondary care

The Review also explored wider issues of how care might be delivered in the future and its implications for training. Locally delivered care will require more doctors trained in broad specialties, including general practice. They will have to be able to manage acute situations in the community with the goal of keeping people out of hospitals as much as possible. Evidence suggests that involving specialists in community care and involving GPs and doctors trained in general areas of a specialty in coordinating hospital and community care lead to:

- improved patient outcomes;
- higher levels of patient and staff satisfaction;
- shorter hospital stays;
- fewer emergency readmissions of acutely ill patients.

Doctors will have to provide emergency and acute care

Although the Review is looking at producing a medical workforce to meet future needs, restructuring training to produce a more broadly trained specialist might ease some of the current workforce pressures. Employers that currently rely on locum doctors and doctors in training to meet service needs, raising patient safety concerns and providing poor levels of supervision, would have other options.

a. The UK government has reported that the crisis in emergency care will likely deepen and may put patients at risk. The GMC has also identified concerns with training in emergency medicine, including poor staffing levels, a lack of supervision by senior doctors and a high and intense workload. Many doctors do not want to train in acute and emergency care because it is perceived as too stressful – ultimately resulting in few doctors able to cover acute care.

b. By training more doctors capable of managing acute and emergency cases, there will be a larger pool of medical staff to cover acute care. This will reduce the stress and intensity of the workload currently experienced by those providing acute care, it will also break the vicious cycle of unattractive areas of medicine failing to recruit staff and so becoming more understaffed, more stressful and more unattractive. We must ensure that caring for acutely ill people is embedded in all specialty training as a core feature and should cover both community and hospital settings.

In summary, the thrust of the report was:

- All doctors must have generalist skills
- All doctors should participate in acute care of patients with their generalist area
- Specialists would still be needed but all specialists should maintain their generalist skills throughout their career and participate in the acute care
- Workforce planning for specialists should be guided by patient needs

The review recommended that implementation of the recommendations must be carefully planned on a UK-wide basis and phased in. This transition period will allow the stability of the overall system to be maintained while reforms are being made. A UK-wide Delivery Group should be formed immediately to oversee the implementation of the recommendations.

Ref: http://www.shapeoftraining.co.uk/home.asp

The implementation of the 'Shape of Training' recommendations

The four UK Departments of Health agreed at the UK Scrutiny Board in February 2014 to form a steering group to consider how to take forward the report's recommendations. The four nation group hosted a series of workshops in September 2014 and reported in November 2014 that after discussion, members agreed a number of draft policy proposals to be presented to the four UK Health Ministers (not presented at the time of going to press February 2015). Timings for further engagement are still to be agreed.

BMA and 15 Trainee organisations have called on for a pause in any implementation of the Shape of Training recommendations. They argue that Shape of Training do not offer the right solutions for patients and could risk all that currently work well in high quality medical education. The key concerns are:

1. Shortening training for training:

The Shape of Training report argues that doctors who are awarded a Certificate of Specialty Training (CST) must be trained to "the same level of competence" as a current Certificate of Completion of Training (CCT) holder. It makes no attempt to explain how doctors can be trained to this skill level in a shorter training programme which has, at the same time, been expanded to include more generalist training.

2. Point of GMC registration

The report recommends moving the point doctors are registered with the General Medical Council (GMC) to the end of medical school. However, unless the length of medical school programmes were extended this would result in the cramming of training and clinical experience currently provided by the F1 year into the undergraduate curriculum. The BMA is not convinced it is possible to produce doctors who are fit to practise under these conditions.

Ref: http://tinyurl.com/l8exps6

End of Interview

The interview may end with the panel asking: Is there anything we could have asked that would influence the selection of candidates?

This is your moment, prepare a short vignette: I am great because... Say I am into forming a patient feedback group—a patient feedback form or whatever unique/special drives you.

Do you have any questions for us?

Always say no thank you.

Say that you have had ample opportunity to discuss the post in depth and you feel clear about the remit of, and challenges in the post.

Post interview

If you don't get the job, you don't get called in first. However, it is vital you should get feedback.

- What was lacking in your CV?
- What else did they want?
- Was there anything you could have done better at the interview?

Lightning Source UK Ltd.
Milton Keynes UK
UKOW04f2348210116

266870UK00001B/223/P